Lecture Notes in Nonlinear Optics: A student's perspective

Instructor:

Mark G. Kuzyk
Department of Physics and Astronomy
Washington State University

Students:

Benjamin R. Anderson
Nathan J. Dawson
Sheng-Ting Hung
Wei Lu
Shiva Ramini
Jennifer L. Schei
Shoresh Shafei
Julian H. Smith
Afsoon Soudi
Szymon Steplewski
Xianjun Ye

December 21, 2013

This book was typeset using LaTeX and set in New Century Schoolbook

Published by NLOsource.com
Pullman, WA 99163

Copyright © 2014 by Mark G. Kuzyk
All rights reserved

Lecture Notes in Nonlinear Optics: A student's perspective

Library of Congress Catalog Number 2013923208
ISBN-13: 978-1494408930
ISBN-10: 1494408937

10 9 8 7 6 5 4 3 2 1

Cover photo courtesy of Fran Morrissey

NLOSource.com

Dedication

We dedicate this book to everyone who is trying to learn nonlinear optics.

Preface

This book grew out of lecture notes from the graduate course in Nonlinear Optics I taught in spring 2010 at Washington State University. Each student took notes for the equivalent of about two book sections, then prepared an electronic version using LaTeX. Each section was criticized by two students who acted as devil's advocates with particular attention to clarity of presentation and accuracy. After making edits, each student presented me with a draft, which I edited and in some cases rewrote. I contributed a chapter, merged all the sections together, and added bridge material when necessary. This book is the result of a process which started on Day 1 of class, and spilled over into the summer.

In addition to this book project, the students gave presentations on modern research that uses nonlinear optics. They were also assigned a class project on two original research topics that I felt were interesting. One project led directly to two publications in *Physics Review A* on cascading and the other one spawned several journal articles on quantum graphs, which are still a topic of research in my group. These additional topics are not included in the lecture notes.

In its present form, this book is a first cut that is still in need of editing – the existing material needs to be expanded, and new material needs to be added. The homework assignments and solutions are incomplete and will be added in a future edition. Nevertheless, the notes serve as a useful reference that contains the fundamentals from the perspective of students who made an effort to distil my lectures into a form that they believe best describes the ideas. In this, I believe they have succeeded. As a result, the pages that follow provide the student with the fundamentals needed to initiate new research in the field and to understand the more advanced topics.

I would like to thank Washington State University for providing a dynamic atmosphere where bright students, faculty, and visitors are encouraged to navigate uncharted waters in search of discovery, fueled solely by passion. I thank my fellow voyagers for sharing their love of learning, mak-

ing lots of work seem effortless and downright fun.
   A pdf file of this book is freely available at:

http://www.nlosource.com/LectureNotesBook.pdf

   Mark G. Kuzyk
   July, 2010
   Pullman, WA

Contents

Dedication   i

Preface   iii

1 Introduction to Nonlinear Optics   1
   1.1 History 1
      1.1.1 Kerr Effect 1
      1.1.2 Two-Photon Absorption 3
      1.1.3 Second Harmonic Generation 4
      1.1.4 Optical Kerr Effect 4
   1.2 Units 7
   1.3 Example: Second Order Susceptibility 8
   1.4 Maxwell's Equations 9
      1.4.1 Electric displacement 11
      1.4.2 The Polarization 13
   1.5 Interaction of Light with Matter 14
   1.6 Goals 18

2 Models of the NLO Response   19
   2.1 Harmonic Oscillator 19
      2.1.1 Linear Harmonic Oscillator 19
      2.1.2 Nonlinear Harmonic Oscillator 21
      2.1.3 Non-Static Harmonic Oscillator 24
   2.2 Macroscopic Propagation 28
   2.3 Response Functions 36
      2.3.1 Time Invariance 37
      2.3.2 Fourier Transforms of Response Functions: Electric Susceptibilities 38
      2.3.3 A Note on Notation and Energy Conservation 39
      2.3.4 Second Order Polarization and Susceptibility 40

  2.3.5 n^{th} Order Polarization and Susceptibility 41
  2.3.6 Properties of Response Functions 41
 2.4 Properties of the Response Function 44
  2.4.1 Kramers-Kronig 44
  2.4.2 Permutation Symmetry 47

3 Nonlinear Wave Equation 53
 3.1 General Technique 53
 3.2 Sum Frequency Generation - Non-Depletion Regime 56
 3.3 Sum Frequency Generation - Small Depletion Regime 60
 3.4 Aside - Physical Interpretation of the Manley-Rowe Equation . . 65
 3.5 Sum Frequency Generation with Depletion of One Input Beam . 66
 3.6 Difference Frequency Generation 70
 3.7 Second Harmonic Generation 72

4 Quantum Theory of Nonlinear Optics 85
 4.1 A Hand-Waving Introduction to Quantum Field Theory 85
  4.1.1 Continuous Theory 85
  4.1.2 Second Quantization 89
  4.1.3 Photon-Molecule Interactions 90
  4.1.4 Transition amplitude in terms of $\vec{E}$ 93
  4.1.5 Stimulated Emission 94
 4.2 Time-Dependent Perturbation Theory 95
  4.2.1 Time-Dependent Perturbation Theory 96
  4.2.2 First-Order Susceptibility 99
  4.2.3 Nonlinear Susceptibilities and Permutation Symmetry . . 101
 4.3 Using Feynman-Like Diagrams 102
  4.3.1 Introduction 102
  4.3.2 Elements of Feynman Diagram 105
  4.3.3 Rules of Feynman Diagram 106
  4.3.4 Example: Using Feynman Diagrams to Evaluate Sum Frequency Generation 108
 4.4 Broadening Mechanisms 111
 4.5 Introduction to Density Matrices 114
  4.5.1 Phenomenological Model of Damping 118
 4.6 Symmetry 123
 4.7 Sum Rules 125
 4.8 Local Field Model 126

5 Using the OKE to Determine Mechanisms — 133

- 5.1 Intensity Dependent Refractive Index 133
  - 5.1.1 Mechanism of $\chi^{(3)}$ 135
- 5.2 Tensor Nature of $\chi^{(3)}_{ijkl}$ 138
- 5.3 Molecular Reorientation 141
  - 5.3.1 Zero Electric field 145
  - 5.3.2 Non-Zero Electric field 146
  - 5.3.3 General Case 147
- 5.4 Measurements of the Intensity-Dependent Refractive Index . . . 149
- 5.5 General Polarization 151
  - 5.5.1 Plane Wave 151
- 5.6 Special Cases 152
  - 5.6.1 Linear Polarization 152
  - 5.6.2 Circular Polarization 152
  - 5.6.3 Elliptical Polarization 153
- 5.7 Mechanisms 154
  - 5.7.1 Electronic Response 154
  - 5.7.2 Divergence Issue 155
  - 5.7.3 One- and Two-Photon States 156

6 Applications — 161

- 6.1 Optical Phase Conjugation 161
- 6.2 Phase Conjugate Mirror 162

7 Appendix - Homework Solutions — 167

Chapter 1

Introduction to Nonlinear Optics

The traditional core physics classes include electrodynamics, classical mechanics, quantum mechanics, and thermal physics. Add nonlinearity to these subjects, and the richness and complexity of all phenomena grows exponentially.

Nonlinear optics is concerned with understanding the behavior of light-matter interactions when the material's response is a nonlinear function of the applied electromagnetic field. In this book, we focus on building a fundamental understating of wave propagation in a nonlinear medium, and the phenomena that result. Such un understanding requires both an understanding of the Nonlinear Maxwell Equations as well as the mechanisms of the nonlinear response of the material at the quantum level.

We begin this chapter with a brief history of nonlinear optics, examples of some of the more common phenomena, and a non-rigorous but physically intuitive treatment of the nonlinear response.

1.1 History

1.1.1 Kerr Effect

The birth of nonlinear optics is often associated with J. Kerr, who observed the change in the refractive index of organic liquids and glasses in the presence of an electric field. [1, 2, 3] Figure 1.1 shows a diagram of the experiment.

Kerr collimated sunlight and passed it through a prism, essentially mak-

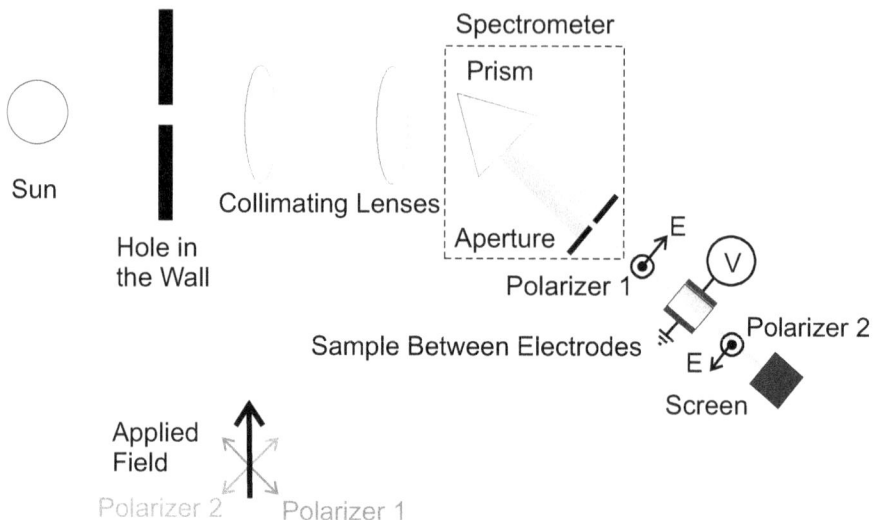

Figure 1.1: Kerr observed the change in transmittance through a sample between crossed polarizers due to an applied voltage. Inset at the bottom left shows the orientations of the polarizers and the applied electric field due to the static voltage.

ing a spectrometer that could be used to vary the color of the light incident on the sample. He placed an isotropic sample between crossed polarizers (i.e. the polarizers' axes are perpendicular to each other), so that no light makes it to the screen. A static electric field is applied $45°$ to the axis of each polarizer as shown in the bottom left portion of Figure 1.1. He found that the transmitted intensity is a quadratic function of the applied voltage.

This phenomena is called the Kerr Effect or the quadratic electrooptic effect; and, originates in a birefringence that is induced by the electric field. The invention of the laser in 1960[4] provided light sources with high-enough electric field strengths to induce the Kerr effect with a second laser in that replaces the applied voltage. This latter phenomena is called the Optical Kerr Effect (OKE). Since the intensity is proportional to the square of the electric field, the OKE is sometimes called the intensity dependent refractive index,

$$n = n_0 + n_2 I, \tag{1.1}$$

where n_0 is the linear refractive index, I is the intensity and n_2 the material-dependent Kerr coefficient.

Lecture Notes in Nonlinear Optics 3

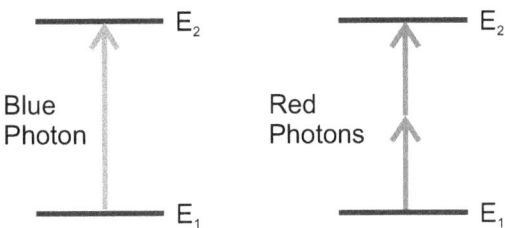

Figure 1.2: (left)A system is excited by a photon if its energy matches the difference in energies between two states. (right) Two-photon absorption results when two photons, each of energy $(E_2 - E_1)/2$, are sequentially absorbed.

1.1.2 Two-Photon Absorption

Two Photon Absorption (TPA) was first predicted by Maria Goeppert-Meyer in 1931,[5] and is characterized by an intensity dependent absorption, of the form,

$$\alpha = \alpha_0 + \alpha_2 I, \tag{1.2}$$

where α_0 is the linear absorption coefficient and α_2 the two-photon absorption coefficient. Since the absorption coefficient is proportional to the imaginary part of the refractive index, a comparison of Equations 1.1 and 1.2 implies that α_2 is proportional to imaginary part of n_2. Thus, both the OKE and TPA are the real and imaginary parts of the same phenomena.

Figure 1.2 can be used as an aid to understand the quantum origin of TPA. A transition can be excited by a photon when the photon energy matches the energy difference between two states (sometimes called the transition energy), $\hbar\omega = E_2 - E_1$, where ω is the frequency of the light. Two-photon absorption results when $\hbar\omega = (E_2 - E_1)/2$, and two photons are sequentially absorbed even when there are no two states that meet the condition $\hbar\omega = E_2 - E_1$.

We can understand TPA as follows. Imagine that the first photon perturbs the electron cloud of a molecule (or any quantum system) so that the molecule is in a superposition of states $|\psi\rangle$ such that the expectation value of the energy is $\hbar\omega = \langle\psi|H|\psi\rangle$. This is called a virtual state, which will be more clearly understood when treated rigourously using perturbation theory. If a second photon interacts with this perturbed state, it can lead to a transition into an excited energy eigenstate. Since the process depends on one photon perturbing the system, and the second photon interacting with the perturbed system, the probability of a double absorption is proportional to the square of the number of photons (the probability of the first absorption times the

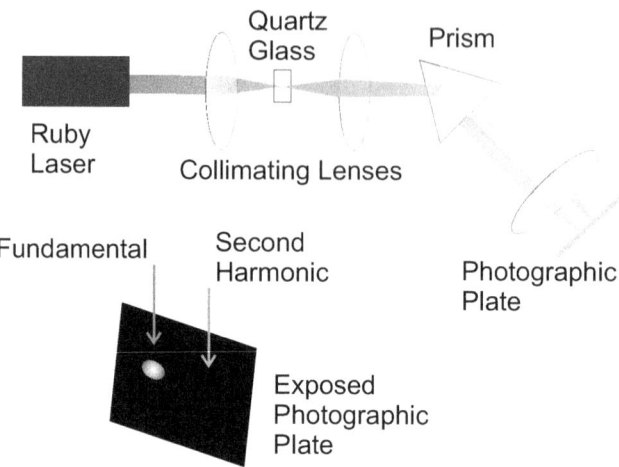

Figure 1.3: The experiment used by Franken and coworkers to demonstrate second harmonic generation. The inset shows an artistic rendition of the photograph that recorded the two beams.

probability of the second absorption), and thus the intensity.

1.1.3 Second Harmonic Generation

Second harmonic generation was the first nonlinear-optical phenomena whose discovery relied on the invention of the laser. P. A. Franken and coworkers focused a pulsed ruby laser into a quartz crystal and showed that light at twice the laser frequency was generated.[6] Figure 1.3 shows the experiment.

A prism was used to separate the the fundamental from the second harmonic, and the two beams were recorded on photographic film. The film's appearance is shown in the inset of Figure 1.3 . Since the second harmonic signal is weak, the spot on the film appeared as a tiny spec. The story is that someone in the publications office thought this spec was a piece of dirt, and air-brushed it away before the journal went to press. So, in the original paper, an arrow points to the location of the second harmonic spot, but the tiny spot is not visible.

1.1.4 Optical Kerr Effect

Three years later, the Optical Kerr Effect was observed by Mayer and Gires.[7] In their experiment, shown in Figure 1.4, a strong light pulse from a ruby laser and a weak probe beam from a xenon flashlamp counterpropagate in

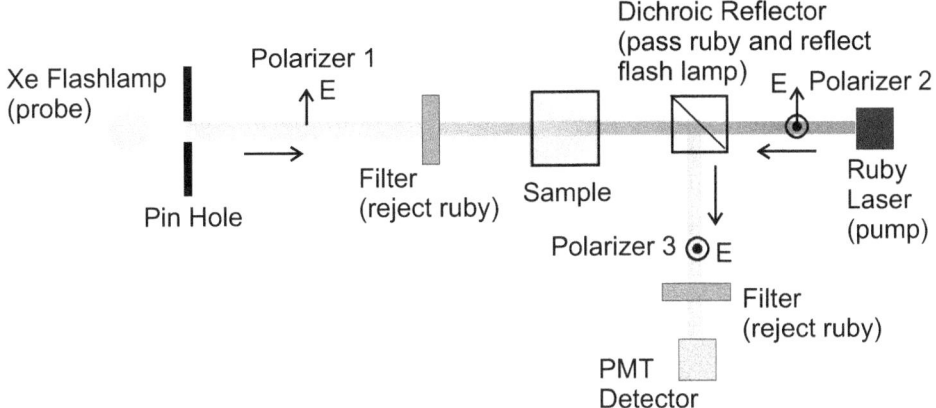

Figure 1.4: In the Optical Kerr Effect, a strong pump laser is polarized $45°$ to the weak probe beam, causing a rotation of its polarization.

the sample. The electric field of the pump and probe beams are polarized $45°$ to each other. The pump beam induces a birefringence in the material that causes a rotation of the polarization of the probe beam. The probe beam is reflected from a dichroic mirror to the detector while allowing the pump beam to pass unaltered. The reflected probe beam passes through polarizer 3, which rejects the original polarization of the probe. Thus, the amount of light reaching the detector is related to the degree of depolarization of the probe beam, which in turn, is proportional to the pump intensity.

While the OKE is similar to the quadratic electrooptic effect, in that a strong electric field rotates the polarization of a weaker probe beam, OKE has many more diverse applications. Interestingly, two-photon absorption also contributes to polarization rotation, as you will show in the exercises below.

Problem 1.1-1(a): A monochromatic plane wave passes through a quartz plate of thickness d with its Poynting Vector perpendicular to the plate. The k-vector is given by $\vec{k} = k\hat{z}$ and the refractive indices along $\hat{x}$ and $\hat{y}$ are n_x and n_y. If the polarization vector of the beam when it enters the sample makes an angle θ with respect to $\hat{y}$, calculate the fraction of the beam intensity that passes through a polarizer that is perpendicular to the light's original polarization.

(b): Now assume that the refractive index is complex so that $n_x \to n_x^R + in_x^I$ and $n_x \to n_y^R + in_y^I$. Recalculate the transmittance (transmitted/insident intensity) through the crossed polarizers.

(c): Next assume that the material is isotropic but that a strong pump beam of intensity I_p is incident at normal incidence and polarized along $\hat{y}$. If $n_y^R = n_0^R + n_{2y}^R I$ and $n_y^I = n_0^I + n_{2y}^I I$, where the refractive index changes only along the direction of the pump beam's polarization, calculate the transmittance of probe beam through the crossed polarizer as a function of the pump intensity.

Homework Hint

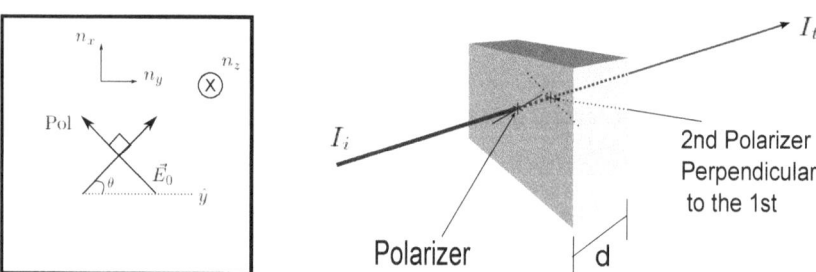

Figure 1.5: Polarizer set perpendicular to the original wave propagation direction (the first defines the polarization of the incident wave.)

Consider a polarizer that is set perpendicular to the polarization of the incident electric field $\vec{E}_0$ as is shown in Figure 1.5. The transmittance is defined as

$$T = \frac{I_t}{I_i}, \qquad (1.3)$$

where the initial intensity is calculated from the electric field,

$$I_i \propto |\vec{E}(z=0)|^2. \tag{1.4}$$

The intensity transmitted through the polarizer is proportional to the square of the transmitted electric field, or

$$I_f \propto |\vec{E}(z=d) \cdot (\cos\theta \hat{x} - \sin\theta \hat{y})|^2, \tag{1.5}$$

where the constant of proportionality is the same for each intensity calculation. The electric field of the wave at any point in the material is given by

$$\vec{E} = E_0 \left(\sin\theta \hat{x} e^{i(kn_x z - \omega t)} + \cos\theta \hat{y} e^{i(kn_y z - \omega t)} \right), \tag{1.6}$$

where $k = \omega/c = 2\pi/\lambda$.

If the refractive index has both a real and an imaginary component, it can be expressed as

$$n = n_R + i n_I, \tag{1.7}$$

where n_R and n_I are the real and imaginary parts of the refractive index. For a pump along $\hat{y}$, assume that

$$n_x = n_{0x} \tag{1.8}$$
$$n_y = n_{0y} + n_{2y} I, \tag{1.9}$$

where Equations 1.8 and 1.9 are substituted into Equation 1.6 with the understanding that all quantities are complex.

1.2 Units

Most of the scientific literature in nonlinear optics has accepted the use of SI units (i.e. meter, kilogram, second, coulomb). However, they lead to ugly equations and mask the beauty of the unification of electric and magnetic fields.

For example, consider Faraday's Law in Gaussian units,

$$\nabla \times \vec{E} = -\frac{1}{c} \frac{\partial \vec{B}}{\partial t}. \tag{1.10}$$

The units of the electric and magnetic fields are the same, placing them on an equal footing.

Coulomb's Law, on the other hand, is given by,

$$\nabla \cdot \vec{E} = 4\pi\rho, \tag{1.11}$$

or for a point charge leads to

$$\vec{E} = \frac{q}{r^2}. \tag{1.12}$$

Note that ϵ_0 and 4π are absent. Also, ϵ_0 does not appear in the induced polarization, which is simply given by

$$P_i = \chi_{ij}^{(1)} E_j, \tag{1.13}$$

where summation convention is implied (i.e. double indices are summed).

There are many conventions for defining the various quantities of nonlinear optics, and this is often a source of much confusion. For example. Boyd's book, "Nonlinear Optics," uses SI units and defines a sinusoidal linearly polarized electric field to be of the form,

$$\vec{E}(t) = E\hat{x}\exp(-i\omega t) + c.c.. \tag{1.14}$$

I prefer the fields to be of the form $\cos\omega t$ rather than $2\cos\omega t$, so I will use the convention,

$$\vec{E}(t) = \frac{1}{2}E\hat{x}\exp(-i\omega t) + c.c. \tag{1.15}$$

Other issues of convention will be explained when they are needed.

1.3 Example: Second Order Susceptibility

It is instructive to give a simple example of the ramifications of a nonlinear polarization before using the full rigor of tensor analysis and nonlocal response. We begin by illustrating the source of second harmonic generation.

Charges within a material are displaced under the influence of an electromagnetic field. The displacement of charges can be represented as a series of electric moments (dipole, quadrupole, etc). Currents can be similarly represented as a series of magnetic moments. The electric dipole field of bound charges is usually (but not always) the largest contribution to the electromagnetic fields. Next in order of importance is the electric quadrupole and magnetic dipole. We use the electric dipole approximation for most of this book and ignore these higher order terms.

In the *linear* dipole approximation, the induced dipole moment per unit volume of material – called the polarization – is proportional to the electric field. The constant of proportionality that relates the applied electric field to the polarization is called the linear susceptibility. If the induced polarization, P, is in the same direction as the applied electric field, E, the linear susceptibility, χ, is a scalar, and we have,

$$P = \chi^{(1)} E. \tag{1.16}$$

It should be stressed that Equation 1.16 is not a fundamental relationship, but rather a model of how charges react to an applied electric field. In the case of Equation 1.16, the model makes an intuitively reasonable assumption that the charges are displaced along the direction of the applied electric field. As we will see later, such models can become more complex when applied to real systems.

If the applied field is sinusoidal, as is the case for an optical field, the induced dipole will oscillate and reradiate the light. In the linear case, the dipole moment oscillates at the same frequency as the incident light.

Now consider a nonlinear material. The polarization will be given by,

$$P = \chi^{(1)} E + \chi^{(2)} E^2 + \ldots, \tag{1.17}$$

where $\chi^{(2)}$ is called the second-order susceptibility. The source of the nonlinear terms ordinates in charges that are bound by an anharmonic potential, which we will study later. If we substitute the scalar form of Equation 1.14 into the second term of Equation 1.16, we get,

$$\begin{aligned} E^2(t) &= \frac{1}{4} E^2 (\exp[-i\omega t] + \exp[+i\omega t])^2 \\ &= \frac{1}{2} E^2 \left(\frac{\exp[-i2\omega t] + \exp[+i2\omega t]}{2} + 1 \right) \end{aligned} \tag{1.18}$$

Equation 1.18 when substituted into Equation 1.17 clearly shows that the polarization oscillates at frequency 2ω, which acts as a source that radiates light at the second harmonic frequency. Also note that the constant offset (the "+1" term) is the source of the phenomena called optical rectification, which results in a static polarization, and therefore a static electric field.

The origin of all the nonlinear-optical processes described in Section 1.1 can be explained using the nonlinear polarization.

1.4 Maxwell's Equations

In this section, we introduce the origin of interactions between light and matter. This requires that we incorporate into Maxwell's equation the effect of light on matter. Motivated by fact that electric fields separate charges in a neutral material, we develop the standard model of describing the charge distribution as a series of moments. These various moments are effected by light depending on how the electric field field varies in from point to point in the material: A uniform electric field induces a dipole moment, a field gradient the quadrupole, etc. This interaction is formulated using the concept of a response function.

We begin by introducing conventions and motivate the use of Gaussian units when expressing Maxwell's Equations due to the symmetry between the electric and magnetic fields that result. We assume non-magnetic materials with no free charges and no free currents. In particular no free charges and free currents implies the absence of sources such as static charge or currents due to batteries.

When light interacts with a material it separates the charges from their equilibrium positions within the constituent molecules or within the lattice of a crystal. Since these charges are intimately associated with the material, they are called *bound charges*, and are "swept under the rug" by the introduction of the electric displacement vector, $\vec{D}$, which has only free electrons as its source..

In Gaussian units, Maxwell's Equations in the absence of free charges and currents are,

$$\nabla \times \vec{E} = -\frac{1}{c}\frac{\partial \vec{B}}{\partial t}, \tag{1.19}$$

$$\nabla \times \vec{B} = \frac{1}{c}\frac{\partial \vec{D}}{\partial t}, \tag{1.20}$$

$$\nabla \cdot \vec{D} = 0, \tag{1.21}$$

$$\nabla \cdot \vec{B} = 0. \tag{1.22}$$

Note that in Gaussian units $\vec{E}$ and $\vec{B}$ have the same units, and are therefore considered on an equal footing.

To gain familiarity with Gaussian units, conisder the divergence of the electric field for a point charge q in vacuum. Gauss' law in Gaussian units is,

$$\nabla \cdot \vec{E} = 4\pi\rho, \tag{1.23}$$

where ρ is the charge density, which when integrated over the volume yields,

$$E = \frac{q}{r^2}. \tag{1.24}$$

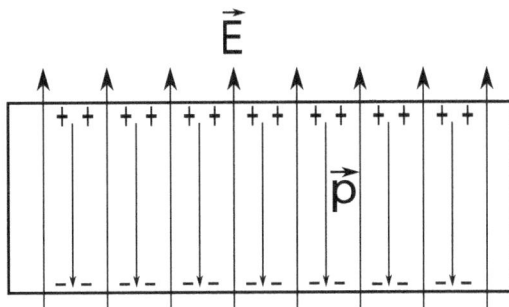

Figure 1.6: Material polarized by an external electric field.

The potential and then the work are given by

$$V = \frac{q}{r} \quad \text{and} \quad U = \frac{q^2}{r}. \tag{1.25}$$

In SI units, ϵ_0, the permittivity of free space, and μ_0, the permeability, yield the speed of light, $c\sqrt{\epsilon_0 \mu_0} = 1$. Indeed, this relationship suggests that light is an electromagnetic wave. However, ϵ_0 and μ are absent in in Gaussian units. Instead, the speed of light appears explicitly.

It is instructive to compare Coulomb's law in SI and Gaussian units. For two electrons separated by an Ångstrom, in SI units the electron charge is $e = 1.6 \times 10^{-19} C$, the radius of orbit in a typical atom is about $r = 1\text{Å} = 10^{-10} m$, and $k = \frac{1}{4\pi\epsilon_0} = 9 \times 10^9 \frac{N \cdot m^2}{C}$, the resulting force is,

$$|F| = 2.304 \times 10^{-8} \ N$$

expressed in Newtons.

In Gaussian units the charge is expressed in statcoulombs (usually called "stat coul" or "esu"), as $e = 4.8 \times 10^{-18} esu$, and the orbital radius is $r = 10^{-8} cm$. In Gaussian units, Equation 1.24 with $\vec{F} = q\vec{E}$ yields,

$$|F| = 2.304 \times 10^{-3} \ dyn,$$

where the force is expressed in dynes. Applying the conversion factor $1 \ dyn = 10^{-5} \ N$, we confirm that the two forces are indeed the same, as expected.

1.4.1 Electric displacement

The electric field in a material is the superposition of the applied electric field, the field to free charge, and the fields due to all of the displaced charges.

The electric field due to the bound charges can be expressed as a series in the moments. Ignoring terms higher order than the dipole term, the electric displaced is given by,

$$\vec{D} = \vec{E} + 4\pi\vec{P} + \ldots \tag{1.26}$$

The higher-order terms will be discussed later. In Equation 1.26, $\vec{E}$ is the total field of all charges(bound and free), $\vec{D}$ is the electric displacement due to free charges and $\vec{P}$ is the electric dipole moment per unit volume, and accounts for the bound charge.

Figure 1.6 shows a material with an applied electric field and the induced charge. The electric field polarizes the material, causing the positive and negative charges to move in opposite directions until they encounter the surface. The net result is a dipole moment. Note that the polarization has the same units as the electric field. The electric field due to the polarization opposes the applied electric field, thus reducing the field strength inside the dielectric.

An electric field gradient results in a different magnitude of force on two neighboring electric charges of the same sign. Thus, two identical neighboring charges will be displaced by differing amounts, resulting in a quadruple movement. Similarly, higher-order field gradients will couple to higher-order moments. To take this into account, the electric displacement can be expressed as a series of moments of the form

$$D_i = E_i + 4\pi P_i + 4\pi \frac{\partial}{\partial x_j} Q_{ij} + 4\pi \frac{\partial^2}{\partial x_j \partial x_k} O_{ijk} + \ldots, \tag{1.27}$$

where Q_{ij} is the quadrupole moment and O_{ijk} is the octupole moment, etc.[1]

A vast number of nonlinear-optical processes can be approximated by the dipole term in Equation 1.27. However, we must be keep in mind that the dipole approximation does not always hold, especially when there are large field gradients, as can be found at interfaces between two materials. Figure 1.7 shoes a plot of the electric field as a function of the surface-normal coordinate z. The abrupt drop at the interface due to the surface charge leads to a large surface gradient, so higher order terms need to be taken into account.

---

[1]Here we use **Summation Convention** where,

$$A_{ij}V_j = \sum_{j=1}^{3} A_{ij}V_j.$$

An index appearing once represents a vector component; an index appearing twice indicates the component that is to be summed over the three cartesian components.

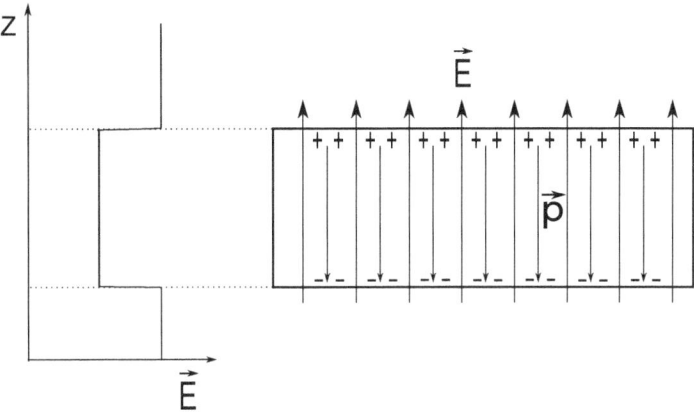

Figure 1.7: Plot of the electric field across the interface of a polarized material.

1.4.2 The Polarization

The electric polarization, $\vec{P}$, is the dipole moment per unit volume, and is a function of the electric field. When the applied electric field strength is much smaller than the electric fields that hold an atom or molecule together, the polarization $\vec{P}(\vec{E})$ can be expanded as a series in the electric field,

$$\vec{P}(\vec{E}) = \vec{P}^{(0)}(\vec{E}) + \vec{P}^{(1)}(\vec{E}) + \vec{P}^{(2)}(\vec{E}) + \ldots, \qquad (1.28)$$

where the first two terms constitute linear optics and the subsequent terms constitute nonlinear optics.

It is instructive to consider the units of the polarization to confirm that they are the same as the units of the electric field. Doing a dimensional analysis, we get,

$$[P] = \left[\frac{\text{dipole}}{\text{volume}}\right] = \left[\frac{\text{charge} \cdot \text{length}}{\text{length}^3}\right] = \left[\frac{\text{charge}}{\text{length}^2}\right].$$

Comparing Equation smith:Punits with the electric field of a point charge given by Equation 1.24 confirms that the polarization is of the same units as the electric and magnetic field. Doing a similar analysis of Equation (1.27) of the higher-order terms, we can determine the units of the quadrupole, octupole, and higher-order moments.

14

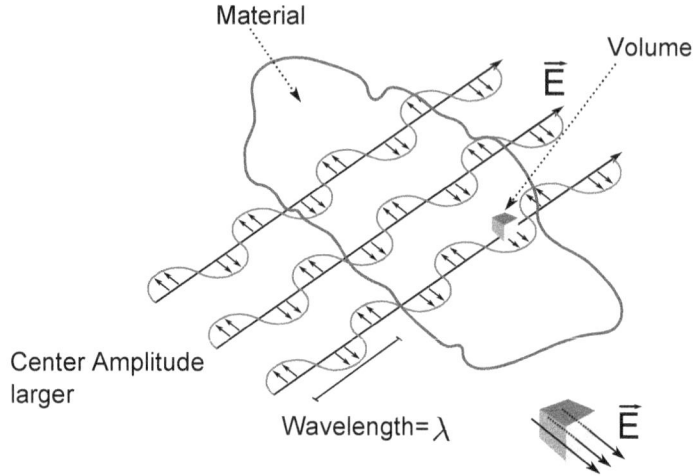

Figure 1.8: Light inducing a polarization inside a material. The light source is a monochromatic plane wave.

1.5 Interaction of Light with Matter

We will consider the most general case of light interacting with matter in the dipole approximation. Figure 1.8 shows the material and the beam of light. We then select a volume, V, that is large enough to contain so many molecules that the material appears homogeneous, but much smaller than a wavelength, so that the electric field is approximated uniform over the volume. The long wavelength approximation that $\lambda \gg \sqrt[3]{V}$ will generally hold in the visible part of the electromagnetic spectrum since light has a wavelength on the order of 1 μm and an atom's size is on the order of 10^{-4} μm.

Now we are ready to formulate the most general theory. Figure 1.8 shows an expanded view of the small volume element in Figure 1.9. Over this volume the electric field, $\vec{E}$, is uniform but varying with time. In this case the most general response is ahs the following properties.

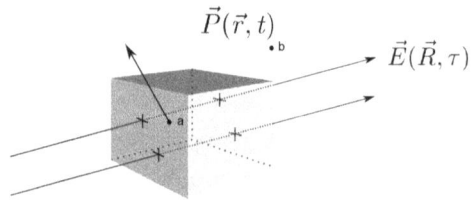

Figure 1.9: A small volume element of material in an electric field

1. $\vec{P}$ is not parallel to $\vec{E}$: We assume that the polarization, $\vec{P}$, can point in any direction, depending on the direction of the electric field and on the material's anisotropy. To account for this, the susceptibility (which relates $\vec{P}$ and $\vec{E}$) must be a tensor.

2. $\vec{P}(\vec{E}) = \underline{Series}$: When the electric field is small, we approximate $\vec{P}$ as a series of the electric field as given by Equation 1.28.

3. The polarization, $\vec{P}$, at point "b" can depend on $\vec{E}$ at point "a": The response may be nonlocal if information about the electric field is transmitted to another point. For example, light absorbed at point "a" can excite an acoustical wave that propagates to "b" and leads to a polarization there. Thus, the polarizability at one point in the material to the electric field everywhere in the material.

4. The polarization at a given time can depend on the electric field in the past: The polarization response can be delayed relative to the electric field. For example, an impulse can start the oscillation of a spring that persists long after the impulse has subsided. Similarly, the charges in a material may come to rest long after the excitation pulse has left the material. Thus, the polarization at any time depends on the electromagnetic history of the material.

After the polarization is expanded as a series in the electric field,

$$\vec{P} = \sum_{n=0}^{\infty} \vec{P}^{(n)}(\mathbf{r}, t) \qquad (1.29)$$

the most general for of the first-order polarization, $\vec{P}^{(1)}(\mathbf{r}, t)$, is given by,

$$P_i^{(1)}(\mathbf{r}, t) = \int_{\substack{all \\ space}} d^3\mathbf{R} \int_{-\infty}^{\infty} d\tau\, T_{ij}(\mathbf{r}, t; \mathbf{R}, \tau) E_j(\mathbf{R}, \tau), \qquad (1.30)$$

where $T_{ij}(\mathbf{r}, t; \mathbf{R}, \tau)$ is the response function, which carries all of the information that relates the applied electric field to the polarization. The tensor form of the response function accounts for anisotropy of the response, and relates the induced polarization along i to the field polarized along j. The non-locality of the response resides in the position dependence; that is, the polarization at coordinate $\mathbf{r}$ is related to the electric field at coordinate $\mathbf{R}$. The integral over all space thus accounts for the influence of every part of the material on the polarization. The polarization at time t in related by the response function to the field at time τ. The integral over time thus takes into account the influence of the the fields in the past.

Clearly, causality must be built into the temporal part of the response function as does the delay to the retarded potentials. Causality will be described later, while the small size of the sample will allow us to ignore retarded potentials.

The second-order polarization is of the form,

$$P_i^{(2)}(\mathbf{r},t) = \int_{\substack{all\\space}} d^3\mathbf{R_1} \int_{\substack{all\\space}} d^3\mathbf{R_2} \int_{-\infty}^{\infty} d\tau_1 \int_{-\infty}^{\infty} d\tau_2 \quad (1.31)$$
$$T_{ijk}(\mathbf{r},t;(\mathbf{R_1},\tau_1),(\mathbf{R_2},\tau_2))E_j(\mathbf{R_1},\tau_1)E_k(\mathbf{R_2},\tau_2).$$

As in the linear case, the polarization at one point in the material is determined by the fields at other points in the material, thus the integration over al space. Again, the polarization at a given time depends on the electric field at previous times, requiring the integral over time. However, the polarization depends on the square of the electric field, so the induced dipole vector depends on the directions of the two electric fields, which is taken into account by the sum over all possible polarizations of both fields. The response function is therefore a third-rank tensor. Boyd described symmetry arguments that can be used to reduce the number of independent tensor components of the response function,[8] thus reducing significantly the complexity of the problem. Details of such symmetry arguments will not be discussed here.

Finally the most general form of the nth-order polarization is expressed as,

$$P_i^{(n)}(\mathbf{r},t) = \int_{\substack{all\\space}} d^3\mathbf{R_1} \ldots \int_{\substack{all\\space}} d^3\mathbf{R_n} \int_{-\infty}^{\infty} d\tau_1 \ldots \int_{-\infty}^{\infty} d\tau_n \quad (1.32)$$
$$T_{ijklm\ldots z}E_j(\mathbf{R_1},\tau_1)\ldots E_z(\mathbf{R_z},\tau_z).$$

Bulk Material

It is instructive to consider the static limit of a spatially uniform electric field. The polarization is then simply given by,

$$P_i = P_i^{(0)} + \chi_{ij}^{(1)}E_j + \chi_{ijk}^{(2)}E_jE_k + \ldots, \quad (1.33)$$

where P_i is the electric dipole moment unit per volume, the first two terms are focus of linear optics, and $\chi^{(n)} = n^{th}$-order susceptibility.

Molecular

For a molecule, the quantum description of volume is not a well defined concept. Thus, it is more appropriate to characterize a molecule in a static electric field by its dipole moment, given by

$$p_i = \mu_i + \alpha_{ij}E_j + \beta_{ijk}E_jE_k + \gamma_{ijkl}E_jE_kE_l, \tag{1.34}$$

where p_i is the dipole moment of the molecule; μ is static dipole moment; α(units of volume) is the polarizability, or otherwise known as the first-order molecular susceptibility; β is the hyperpolarizability, also known as the second-order molecular susceptibility; and γ is the second hyperpolarizability, or the third order molecular susceptibility.

Calculating the Nonlinear Susceptibilities

In light of Equation 1.33, $\chi_{ijk...}^{(n)}$ can be calculated from a model of $\vec{p}(\vec{E})$ by differentiation,

$$\chi_{ijk...}^{(n)} = \frac{N}{n_x!n_y!n_z!} \frac{\partial^n p_i}{\partial E_j \partial E_k ...}\bigg|_{\vec{E}=0} \tag{1.35}$$

where n_i represents the number of fields polarized along the ith cartesian direction and N is the number density of molecules. A similar differentiation can be applied to the molecular dipole moment given by Equation 1.34 to get the molecular susceptibilities.

Example:

(a) A model of the polarization gives,

$$P_x = aE_x^2. \tag{1.36}$$

Determine $\chi_{xx}^{(1)}$ and $\chi_{xxx}^{(2)}$.

(b) Given the polarization

$$P_x = aE_x + bE_yE_z^2, \tag{1.37}$$

find $\chi_{xy}^{(1)}$ and $\chi_{xyzz}^{(3)}$.

Solution:

(a) Taking the derivative as with respect to the field and evaluating it at zero field

$$\chi_{xx}^{(1)} = \frac{1}{1!}\frac{\partial P_x}{\partial E_x}\bigg|_0 = 2aE_x|_0 = 0. \quad (1.38)$$

$$\chi_{xxx}^{(2)} = \frac{1}{2}\frac{\partial^2 P_x}{\partial E_x^2}\bigg|_0 = a. \quad (1.39)$$

(b) Here we apply the same method as in part (a)

$$\chi_{xy}^{(1)} = \frac{1}{1!}\frac{\partial P_x}{\partial E_y}\bigg|_{\vec{E}=0} \quad (1.40)$$

and taking the derivative

$$\chi_{xy}^{(1)} = bE_z^2\big|_{\vec{E}=0}, \quad (1.41)$$

We get,

$$\chi_{xy}^{(1)} = 0. \quad (1.42)$$

For $\chi_{xyzz}^{(3)}$ we can apply the same method, getting

$$\chi_{xyzz}^{(3)} = \frac{1}{1!}\frac{1}{2!}\frac{\partial^3 P_x}{\partial E_y \partial E_z \partial E_z}\bigg|_{\vec{E}=0} = b. \quad (1.43)$$

1.6 Goals

Nonlinear optics encompasses a broad range of phenomena. The goal of this course is to build a fundamental understanding that can be applied to both the large body of what is known as well as to provide the tools for approaching the unknown. In your studies, keep the following questions in in mind:

- Are the nonlinear optical process related to each each other, and if so, how?

- What are the underlying microscopic mechanisms?

- How are the microscopic mechanisms related to macroscopic observations?

- What governs wave propagation in the nonlinear regime?

- How can nonlinear optics be applied?

Chapter 2

Models of the NLO Response

In this chapter, we develop the classical nonlinear harmonic oscillator model of the nonlinear-optical response of a material, show how the nonlinear-optical susceptibility in the frequency domain is related to the response function, and show how symmetry arguments can be used to reduce the number of independent tensor components.

2.1 Harmonic Oscillator

In this section, we begin by developing a model of the static response of a linear harmonic oscillator and generalize it to the nonlinear case to show how the nonlinear susceptibility is related to the linear and nonlinear spring constant. The model is subsequently applied to a nonlinear spring in a harmonic electric field to develop an understanding of the dependence of the nonlinear-optical susceptibility on the frequency of the electric field.

2.1.1 Linear Harmonic Oscillator

Consider as an example a charge on a spring as shown in Figure 2.1. For the static, one dimensional harmonic oscillator model, we apply a static electric field, $\vec{E}$, which causes the spring to stretch.

The applied electric field can be so large that it becomes unraveled and leads to anharmonic terms. But, we assume the field is small enough to keep the spring intact. This is analogous to a molecule where the applied field is large enough to strongly displace the electrons from their equilibrium positions, but not strong enough to ionize the molecule.

Figure 2.1.1 shows a free body diagram illustrating the forces acting on

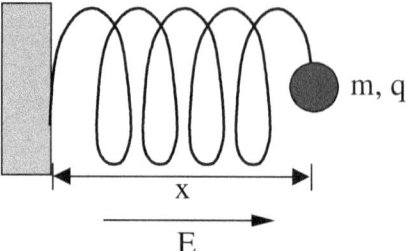

Figure 2.1: A charge on a spring with mass, m, and charge, q, at a distance, x, from the origin. There is an applied external electric field, E.

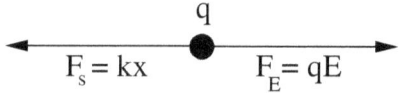

Figure 2.2: Free body diagram for a charge attached to a spring with an applied electric field.

the charge. The force on the charge due to the electric field is balanced by an opposing force due to the tension of the spring.

At equilibrium, the forces are equal to each other, that is,

$$F_s = F_E. \tag{2.1}$$

Substituting the forces into Equation 2.1 yields,

$$kx = qE. \tag{2.2}$$

Solving for x,

$$x = \frac{q}{k}E, \tag{2.3}$$

and using the definition of the dipole moment,

$$p = xq, \tag{2.4}$$

substitution of Equation 2.3 into Equation 2.4 yields

$$p = \frac{q^2}{k}E. \tag{2.5}$$

The static linear susceptibility is defined as,

$$\alpha = \left.\frac{\partial p}{\partial E}\right|_{E=0}, \tag{2.6}$$

which applied to Equation 2.5 yields the linear susceptibility,

$$\alpha = \frac{q^2}{k}. \qquad (2.7)$$

When k is small, the charge is loosely bound and large displacements result from small applied fields. When q is large, the linear susceptibility, or polarizability, is large because the applied force scales in proportion to q.

For N springs per unit volume, as shown in Figure 2.3, we get

$$\chi^{(1)} = N\alpha. \qquad (2.8)$$

Substituting Equation 2.7 into Equation 2.8, we get

$$\chi^{(1)} = N\frac{q^2}{k}. \qquad (2.9)$$

Note that Equation 2.9 explicitly assumes that the springs are non-interacting, and this approximation is often called the dilute gas approximation.

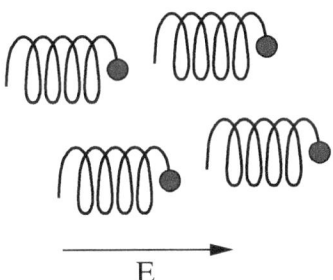

Figure 2.3: A system of N charges on springs with an applied electric field.

The susceptibility is linear with respect to N for low concentrations. However, as N becomes sufficiently large, the susceptibility diverges from linearity (as depicted in Figure 2.4). In systems such as dye-doped polymers, deviations from linearity is a sign of molecular aggregation.

2.1.2 Nonlinear Harmonic Oscillator

Now we will consider the nonlinear harmonic oscillator model. In this system, the potential has a quadratic dependence on the displacement, x. Real systems deviate from the quadratic potential of a simple harmonic oscillator when the amplitude of x gets large (Figure 2.5). In this limit, we can approx-

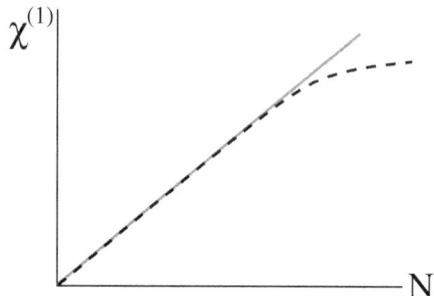

Figure 2.4: Susceptibility as a function of N, the number of molecules. With large N, the susceptibility diverges from linearity because the molecules will begin to aggregate.

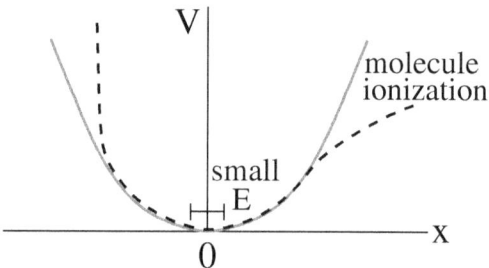

Figure 2.5: The electric potential as a function displacement. As x gets large, the potential no longer follows a quadratic equation and the molecule will ionize beyond some critical displacement. We will consider small displacements when E is small.

imate x is small when the electric field, E, is small. The spring force is,

$$F_s = k^{(1)}x + k^{(2)}x^2. \tag{2.10}$$

In equilibrium, the spring force is equal to the electric force so that,

$$k^{(1)}x + k^{(2)}x^2 = qE. \tag{2.11}$$

Rearranging Equation 2.11 yields,

$$x^2 + \frac{k^{(1)}}{k^{(2)}}x - \frac{q}{k^{(2)}}E = 0. \tag{2.12}$$

Solving for x, we get

$$x = \frac{-k^{(1)}}{2k^{(2)}} \pm \frac{1}{2}\sqrt{\left(\frac{k^{(1)}}{k^{(2)}}\right)^2 + \frac{4q}{k^{(2)}}E}, \tag{2.13}$$

and factoring out $\left(\frac{k^{(1)}}{k^{(2)}}\right)^2$,

$$x = \frac{-k^{(1)}}{2k^{(2)}}\left[1 \mp \sqrt{1 + \frac{4q}{k^{(2)}}\left(\frac{k^{(2)}}{k^{(1)}}\right)^2 E}\right]. \tag{2.14}$$

For a small electric field, we can expand the argument and get

$$x = \sqrt{1 + \frac{4q}{k^{(2)}}\left(\frac{k^{(2)}}{k^{(1)}}\right)^2 E} = 1 + \frac{1}{2}\frac{4q}{k^{(2)}}\left(\frac{k^{(2)}}{k^{(1)}}\right)^2 E + \ldots. \tag{2.15}$$

Using the series expansion of the square root, and keeping only the negative root because it yields the correct linear result, we get,

$$\begin{aligned}
x &= \frac{k^{(1)}}{2k^{(2)}}\frac{1}{2}\frac{4q}{k^{(2)}}\left(\frac{k^{(2)}}{k^{(1)}}\right)^2 E + \ldots \\
x &= \frac{q}{k^{(1)}}E + \ldots. \tag{2.16}
\end{aligned}$$

Note that this is the same result as the linear case where $\alpha = \frac{q^2}{k}$.

Expanding the position given by Equation 2.16 to include higher-order terms, and using Equation 1.35, the susceptibilities are,

$$\chi^{(1)} = N\alpha \tag{2.17}$$
$$= N\frac{q^2}{k^{(1)}} \tag{2.18}$$
$$\text{and} \tag{2.19}$$
$$\chi^{(2)} = \frac{1}{2}\frac{\partial^2 P}{\partial E^2}\bigg|_{E=0} \tag{2.20}$$
$$= \frac{-Nq^3}{(k^{(1)})^2}\frac{k^{(2)}}{k^{(1)}}, \tag{2.21}$$

where N is the number density. Note that as $k^{(2)} \to 0$ then $\chi^{(2)} \to 0$, in agreement with the fact that the nonlinearity originates in $k^{(2)}$. The third-order susceptibility depends quadratically on $k^{(2)}$ and is of the form,

$$\begin{aligned}
\chi^{(3)} &= \frac{1}{6}\frac{\partial^3 P}{\partial E^3}\bigg|_{E=0} \\
&= \frac{2Nq^4}{(k^{(1)})^3}\left(\frac{k^{(2)}}{k^{(1)}}\right)^2. \tag{2.22}
\end{aligned}$$

It is useful to review the units for the susceptibilities in Gaussian units. They are:

$$\alpha = [\text{cm}^3] \qquad \chi^{(1)} = [\text{dimensionless}],$$

$$\beta = \left[\frac{\text{cm}^5}{\text{esu}}\right] \qquad \chi^{(2)} = \left[\frac{\text{cm}}{\text{stat volt}}\right] = \left[\frac{\text{cm}^2}{\text{esu}}\right] = \left[\frac{\text{cm}^3}{\text{erg}}\right]^{\frac{1}{2}},$$

$$\gamma = \left[\frac{\text{cm}^6}{\text{erg}}\right] \qquad \chi^{(3)} = \left[\frac{\text{cm}^5}{\text{erg}}\right].$$

Orders of magnitude ranges for typical molecular susceptibilities are:

$$\alpha \qquad 10^{-21} - 10^{-22} \text{cm}^3,$$

$$\beta \qquad 10^{-27} - 10^{-30} \frac{\text{cm}^5}{\text{esu}},$$

$$\gamma \qquad 10^{-36} \frac{\text{cm}^6}{\text{erg}}.$$

In order to observe these nonlinear effects, the field strengths must be large enough to induce a displacement of charge but small enough so that the molecules do not ionize. For example, using typical field strengths for hydrogen atoms, the field strength must be on the order

$$E \quad << \quad 10^{11} \frac{\text{V}}{\text{m}},$$

$$E \quad << \quad 10^{7} \frac{\text{stat volt}}{\text{cm}}.$$

2.1.3 Non-Static Harmonic Oscillator

For the non-static 1D linear harmonic oscillator model, the equation of motion of an electron derived from Newton's Laws, is,

$$\begin{aligned} F &= m\ddot{x} \\ &= -m\omega_0^2 x - m\Gamma\dot{x} - eE_0\cos(\omega t), \end{aligned} \quad (2.23)$$

where $-m\omega_0^2 x$ is the restoring force, $m\Gamma\dot{x}$ is the damping force, and $eE_0\cos(\omega t)$ is the driving force on the electron. For the nonlinear case, the force includes higher-order terms in the potential,

$$F = -m\omega_0^2 x - m\Gamma\dot{x} - eE_0\cos(\omega t) - \left(\xi^{(2)} x^2 + \xi^{(3)} x^3 + \ldots\right). \quad (2.24)$$

Lecture Notes in Nonlinear Optics

In the linear case $\xi^{(n)} = 0$ for $n > 2$.

A harmonic field of the form,

$$E = E_0 \cos(\omega t), \quad (2.25)$$

can be expressed in complex form,

$$E = \frac{E_0}{2} e^{-i\omega t} + \text{c.c.}, \quad (2.26)$$

where the electric field is the real part

$$E = \text{Re}\left[E_0 e^{i\omega t}\right]. \quad (2.27)$$

Once transients have decayed, it is commonly assumed that the electron oscillates at the driving frequency so that a trial solution of the form,

$$x = \frac{A}{2} e^{-i\omega t} + \frac{A^*}{2} e^{i\omega t}, \quad (2.28)$$

is used. The phase of A takes into account the phase shift between the driving field and the displacement of the electron. Alternatively, we can express the position of the electron by

$$x = x_0 \cos(\omega t + \phi) \quad (2.29)$$

or using a trigonometric identity,

$$x = x_0 \left(\cos(\omega t)\cos(\phi) - \sin(\omega t)\sin(\phi)\right). \quad (2.30)$$

For the linear case $\xi^{(n)} = 0$ for $n > 2$, we show that these two representations are equivalent with

$$A = A_R + iA_I, \quad (2.31)$$

as follows.

The first order solution is of the form,

$$
\begin{align}
x^{(1)} &= \text{Re}\left[\frac{A}{2} e^{-i\omega t}\right] & (2.32) \\
&= \text{Re}\left[\frac{A_R + A_I i}{2} (\cos(\omega t) - i\sin(\omega t))\right] & (2.33) \\
&= \text{Re}[A_R \cos(\omega t) + A_I \sin(\omega t) + i * \text{stuff}] & (2.34) \\
&= A_R \cos(\omega t) + A_I \sin(\omega t), & (2.35)
\end{align}
$$

where $A_R \cos(\omega t)$ is in phase with the field and $A_I \sin(\omega t)$ is $\frac{\pi}{2}$ out of phase with the field. Solving for the components,

$$A_R = x_0 \cos(\phi), \tag{2.36}$$

$$A_I = -x_0 \sin(\phi), \tag{2.37}$$

$$\phi = -\tan^{-1}\left(\frac{A_I}{A_R}\right), \tag{2.38}$$

and

$$x_0 = \sqrt{(A_R^2 + A_I^2)}. \tag{2.39}$$

Equations 2.36 through 2.39 show that one can transform between (A_r, A_I) to (x_0, ϕ).

Substituting the ansatz given by Equation 2.28 into the linear equation of motion given by Equation 2.23 and solving for A, we get

$$A = \frac{-eE_0}{m} \frac{1}{\omega_0^2 - 2i\gamma\omega - \omega^2}. \tag{2.40}$$

Multiplying the numerator and denominator by $((\omega_0^2 - \omega^2) + 2i\gamma\omega)$, A becomes,

$$A = \frac{-eE_0}{m} \frac{(\omega_0^2 - \omega^2) + 2i\gamma\omega}{(\omega_0^2 - \omega^2)^2 + 4\gamma^2\omega^2}. \tag{2.41}$$

The dipole moment can be written as,

$$P^{(1)} = -eNx^{(1)}. \tag{2.42}$$

Substituting Equation 2.28 into Equation 2.23, with the help of Equation 2.41 we get,

$$P^{(1)} = \frac{e^2 E_0 N}{2m} \frac{1}{D(\omega)} e^{-i\omega t} + \text{c.c.} \tag{2.43}$$

But,

$$P^{(1)} = \frac{P_\omega^{(1)}}{2} e^{-i\omega t} + \frac{P_{-\omega}^{(1)}}{2} e^{i\omega t}, \tag{2.44}$$

where $P_\omega^* = P_{-\omega}$. These are the Fourier components at ω and $-\omega$. Then the dipole moment can be written as

$$P^{(1)} = \text{Re}\left[\chi^{(1)} E\right]$$

$$= \frac{\chi^{(1)}}{2} E_0 e^{-i\omega t} + \text{c.c.} \tag{2.45}$$

Setting Equations 2.43 and 2.45 equal to each other and taking the Fourier component at ω, we get

$$\begin{aligned}\chi^{(1)} &= \frac{e^2 N}{m} \frac{(\omega_0^2 - \omega^2) + 2i\Gamma\omega}{(\omega_0^2 - \omega^2)^2 + 4\Gamma^2\omega^2} \\ &= \frac{e^2 N}{m} \frac{1}{D(\omega)}.\end{aligned} \quad (2.46)$$

Figure 2.6 illustrates the real and imaginary components of $\chi^{(1)}$. The real part represents the in-phase component while the imaginary part is the $\frac{\pi}{2}$ out of phase component of the induced dipole moment with the driving field.

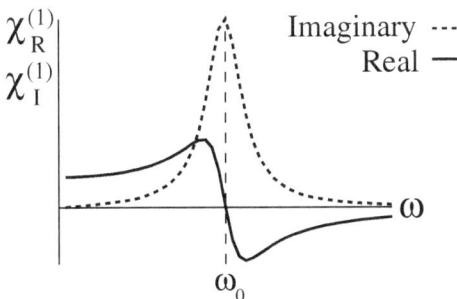

Figure 2.6: The real and imaginary parts of the susceptibility of a harmonic oscillator.

In the second-order non-linear case, $\xi^{(2)} \neq 0$ but $\xi^{(n)} = 0$ for $n > 2$. Using classical perturbation theory, we assume that the solutions are of the form $x = \lambda x^{(1)} + \lambda^2 x^{(2)}$. For small fields, $E = \lambda E_0 \cos(\omega t)$. We already solved this equation to first order where we calculated $x^{(1)}$. For order $x^{(2)}$ (calculated by considering terms of order λ) the force equation gives

$$m\left[\ddot{x}^{(2)} + 2\Gamma\dot{x}^{(2)} + \omega_0^2 x^{(2)} - \xi^{(2)}(x^{(1)})^2\right] = 0, \quad (2.47)$$

where $\xi^{(2)}(x^{(1)})^2$ acts as the driving term. Thus, we must first solve for $x^{(1)}$ before we can solve for $x^{(2)}$. For the general case of 2 fields,

$$E = \frac{E_1}{2} e^{-i\omega_1 t} + \frac{E_2}{2} e^{-i\omega_2 t} + \text{c.c}, \quad (2.48)$$

and

$$x^{(1)} = x^{(1)}_{\omega_1} + x^{(1)}_{\omega_2}. \quad (2.49)$$

Then,

$$x^{(1)} = \frac{A_1}{2}e^{-i\omega_1 t} + \frac{A_1^*}{2}e^{i\omega_1 t} + \frac{A_2}{2}e^{-i\omega_2 t} + \frac{A_2^*}{2}e^{i\omega_2 t} \quad (2.50)$$

Substituting Equation 2.50 into Equation 2.47, we get terms of the form $e^{-i2\omega_1 t}$, $e^{-i2\omega_2 t}$, $e^{-i(\omega_1 \pm \omega_2)t}$, and a constant in time. Thus, we solve the problem by projecting out each of these Fourier components. To illustrate, we project out Fourier component at frequency $2\omega_1$,

$$\int_{-\inf}^{\inf} d\omega \left(m\left[\ddot{x}^{(2)} + 2\Gamma \dot{x}^{(2)} + \omega_0^2 x^{(2)} - \xi^{(2)}(x^{(1)})^2 \right] = 0 \right) \cos(2\omega_1 t), \quad (2.51)$$

which yields,

$$m\left[\ddot{x}^{(2)}_{2\omega_1} + 2\Gamma \dot{x}^{(2)}_{2\omega_1} + \omega_0^2 x^{(2)}_{2\omega_1} \right] = \xi^{(2)} \frac{m}{2}\left[\frac{A^2}{2}e^{-2i\omega_1 t} + \text{c.c} \right], \quad (2.52)$$

where we have used the fact that $x^{(2)} = x^{(2)}_{2\omega_1} + x^{(2)}_{2\omega_2} + x^{(2)}_{\omega_1 \pm \omega_2} + \ldots$. Note that Equation 2.52 is the same form as Equation 2.23. Solving Equation 2.52 using the same approach that is used to solve Equation 2.23 yields the displacement,

$$x^{(2)}_{2\omega_1} = \xi^{(2)} \frac{(\frac{e}{m})^2 E_1^2}{D^2(\omega_1)D(2\omega_1)}, \quad (2.53)$$

where E_1 is the amplitude of the field at ω_1. To solve for the response at the sum frequency, we project out the $\omega_1 + \omega_2$ Fourier component of Equation 2.47. This yields,

$$x^{(2)}_{\omega_1+\omega_2} = -2\xi^{(2)} \frac{(\frac{e}{m})^2 E_1 E_2}{D(\omega_1+\omega_2)D(\omega_1)D(\omega_2)}. \quad (2.54)$$

We can use a similar approach to get the response at other frequencies. The nonlinear susceptibilities can be calculated from Equations 2.53 and 2.54 by recognizing that the dipole moment is given by $p = -ex$ with a polarization given by $P = Np$, and then applying Equation 1.35.

2.2 Macroscopic Propagation

In the previous sections, we developed intuition about the bulk nonlinear processes and then studied the nonlinear oscillator to learn about the microscopic models at the molecular level. Next we considered the macroscopic behavior of the non-interacting gas model, where we related $\chi^{(n)}$ to α, β and

γ. Here we study the bulk response independent of the the source of the response. First we consider macroscopic propagation crudely to gain physical insight, then we study it in more detail using the full mathematical formalism.

Understanding $\chi^{(2)}(\omega_1,\omega_2)$ enables us to describe a number of processes like second harmonic generation(SHG), sum frequency generation(SFG) and optical rectification(OR). On the other hand, $\chi^{(2)}$ is a function of frequencies ω_1 and ω_2, which provides significant information about the characteristics of the system.

Two frequencies are sufficient to understand all the characteristics of $\chi^{(2)}$ because $P^{(2)}$ is proportional to the product of two fields like $E^{\omega_1}E^{\omega_2}$, $E^{\omega_1}E^{\omega_3}$ and so forth. Thus the fields in $\chi^{(2)}$ always appear in pairs and any pair of fields are represented by $\chi^{(2)}(\omega_1,\omega_2)$. Similarly, n frequencies are required to describe the nth order nonlinear response.

As mentioned above, studying the resonance frequencies of a system is a general approach to explore its characteristics. The resonance frequency is defined as a frequency for which the real part of the responses (e.g. $\chi^{(1)}$) changes sign and the imaginary part is characterized by a peak. $\chi^{(2)}$ and the higher order processes show more complete resonance structures. To describe the nonlinearities of a system, one needs to characterize all the frequencies. Quantities such as $\chi^{(1)}$ and its resonance frequencies provide some information about the system. $\chi^{(2)}$ can give more information because it is sensitive to different symmetries than is $\chi^{(1)}$. Therefore resonances in $\chi^{(2)}$ may indicate states that can't be seen as features in the linear absorption spectrum. As an example, nonlinear microscopy is used in biological systems to bring out certain structures that are associated with orientation order and two-photon states.

To begin, let's look at the resonance frequency for the simple case of $\chi^{(1)}$. In Figure 2.7, we plot the imaginary and real parts of $\chi^{(1)}$ versus frequency ω. The natural (resonant) frequency of the oscillation of the system is ω_0.

For $\chi^{(2)}$ we have

$$\chi^{(2)}(\omega_1,\omega_2) = \frac{1}{\mathscr{D}} \frac{\partial^2 (qNx_i)}{\partial E_j^{\omega_1} \partial E_k^{\omega_2}}\bigg|_{\mathbf{E}=0}, \qquad (2.55)$$

where N is the number density of the molecules, qx is dipole moment, qNx is the polarization and $\mathscr{D}$ is the degeneracy factor, which should not be confused with the energy denominator $D(\omega_1,\omega_2)$. In previous cases, $\mathscr{D}$ depended on the tensor components of the frequencies of the system, but for $\chi^{(2)}$, it also depends on ω_1 and ω_2. The surface plot of $\chi^{(2)}$ as a function of frequencies

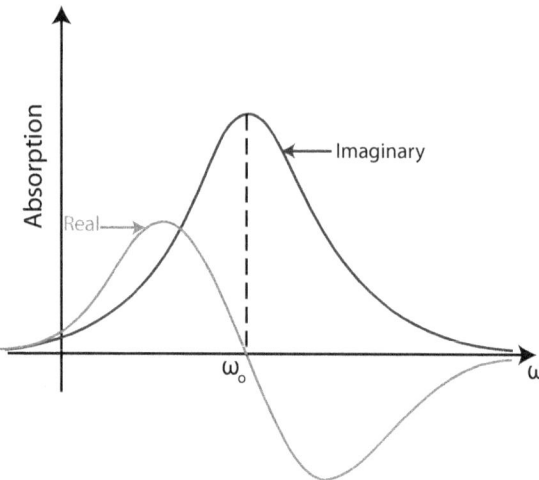

Figure 2.7: Plot of the real and imaginary parts of $\chi^{(1)}$. ω_0 is the resonant frequency of the system.

becomes complicated since the energy denominators in $\chi^{(2)}$ include different combination of frequencies. Hence it is more interesting to visual the resonances with a two-dimensional plot of the second-order susceptibility as a function of the frequencies ω_1 and ω_2. In the ω_2-ω_1 plane, one can determine where the resonances occur. Since the energy denominators of $\chi^{(2)}$ are of the form $D(\omega_1)$, $D(\omega_2)$, $D(2\omega_1)$ and $D(2\omega_2)$ the system has resonances when $\omega_1 = \omega_0$, $\omega_2 = \omega_0$, $\omega_1 = \omega_0/2$, $\omega_2 = \omega_0/2$, and $\omega_2 \pm \omega_0 = \omega_0$. These resonances are shown in Figure 2.8, where the various lines shows the position of the resonances.

A simple macroscopic model can be used to explain the concept of refractive index. In Figure 2.9, plane waves are incident from the left on the sample. This external electric field, E^ω, induces dipoles in the molecule, which consequently generate dipole fields. The superposition of the incident and induced electric fields propagates through the material then exits to the right. The induced dipoles oscillate at the frequency of the plane wave. Assuming the plane waves and the sample, which is made of dipoles, are infinite in the transverse direction, the superposition of the induced dipole fields leads to planes of constant phase that are parallel to the incoming wave's phase fronts. The wavelength of the reflected dipole fields propagating in the opposite direction of the incident wave to the left (Figure 2.9) is also the same as of the plane wave.

The superposition of the incident and dipole fields inside the material

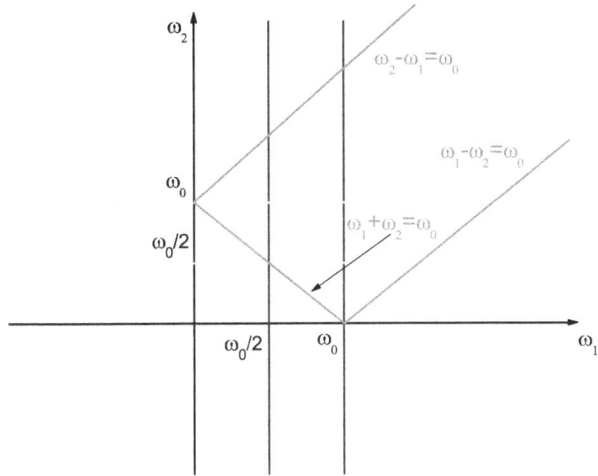

Figure 2.8: Map of resonances of $\chi^{(2)}$

leads to a phase shift relative to the incident wave and of lower phase velocity in the sample. The refractive index quantifies the reduced phase velocity inside the material, hence, the wave propagates more slowly, and the peaks are closer together. Outside, the net effect is a phase shift of the transmitted wave.

When a nonlinear material is placed in the path of a beam of light of frequency of ω (for now, consider only a second order nonlinearity, $\chi^{(2)} \neq 0$), the dipoles will oscillate at frequency 2ω and if the field is of the form $E^{\omega_1} + E^{\omega_2}$, then the frequencies of oscillation are 0, $2\omega_1$, $2\omega_2$, $\omega_1 + \omega_2$ and $|\omega_1 - \omega_2|$. Therefore, outside the sample we observe fields with other frequencies. For $\chi^{(1)}$, radiation with frequencies ω_1 and ω_2 are observed.

Next, we use Maxwell's equations to rigorously solve for nonlinear wave propagation. First, we find a relationship for ϵ_0 when the nonlinear response vanished and then, we explore how the refractive indices are generalized for systems in which $\chi^{(2)}$ and $\chi^{(3)}$ are dominant.

In Gaussian units,

$$\mathbf{D} = \mathbf{E} + 4\pi \mathbf{P} = \overleftrightarrow{\epsilon}\, \mathbf{E}, \tag{2.56}$$

where $\overleftrightarrow{\epsilon}$ is a rank two tensor. Note that in Gaussian units, The permittivity of free space, ϵ_0, is absent. When the nonlinear response vanishes,

$$\mathbf{P} = \chi^{(1)} \mathbf{E}. \tag{2.57}$$

Then

$$\mathbf{D} = \mathbf{E} + 4\pi \chi^{(1)} \mathbf{E} = \left(1 + 4\pi \chi^{(1)}\right) \mathbf{E}, \tag{2.58}$$

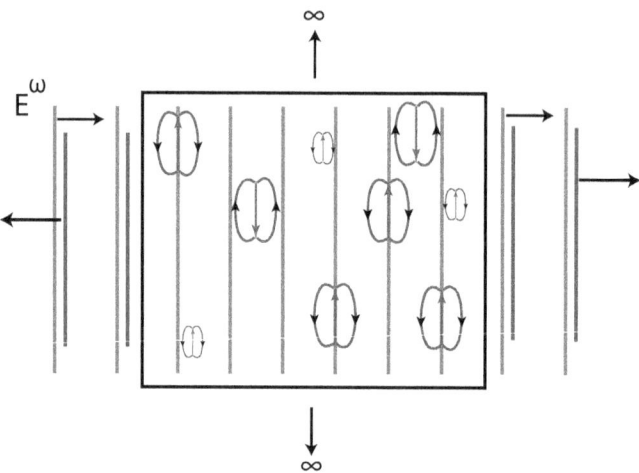

Figure 2.9: Illustration of the macroscopic model that defines refractive index. Longer (red) lines represent phase fronts of the external electric field in and outside the material and the shorter (dark blue) lines represent the electric field generated by the induced dipoles fields (arrows) inside the sample. The superposition of the induced electric dipole fields and plane waves inside the sample leads to a plane wave with a decreased phase velocity, which defines the concept of the refractive index.

whence
$$\epsilon_0 = 1 + 4\pi\chi^{(1)}, \tag{2.59}$$

or in tensor form,
$$\epsilon_{ij} = \delta_{ij} + 4\pi\chi^{(1)}_{ij}. \tag{2.60}$$

ϵ_0 is the linear dielectric constant of the material. The method used to find ϵ_0, through Equation 2.59 provides a simple approach to quantifying nonlinear wave propagation without solving Maxwell's equations.

To obtain the wave equation for the regions inside the material, we apply Faraday's and Ampere's laws. Based on Faraday's law

$$\nabla \times \mathbf{E} = -\frac{1}{c}\frac{\partial \mathbf{B}}{\partial t}, \tag{2.61}$$

and taking the curl of Equation 2.61, leads to

$$\nabla \times \nabla \times \mathbf{E} = -\frac{1}{c}\frac{\partial}{\partial t}\nabla \times \mathbf{B}. \tag{2.62}$$

Lecture Notes in Nonlinear Optics

Inside the sample, where there are no free currents – as, for example, one finds when an external source is applied – Ampere's law gives,

$$\nabla \times \mathbf{B} = \frac{1}{c}\frac{\partial \mathbf{D}}{\partial t}, \tag{2.63}$$

where **H** is equivalent to **B** when the material is nonmagnetic and the magnetic susceptibility is equal to unity. Inserting Equation 2.63 in Equation 2.62 yields

$$\nabla \times \nabla \times \mathbf{E} + \frac{\epsilon_0}{c^2}\frac{\partial^2 \mathbf{E}}{\partial t^2} = -\frac{4\pi}{c^2}\frac{\partial^2 \mathbf{P}_{NL}}{\partial t^2}, \tag{2.64}$$

where the linear polarization, $4\pi\chi^{(1)}\mathbf{E}$, is absorbed into the second term of Equation 2.64. The nonlinear polarization, $\mathbf{P}_{NL}$, is given by

$$\mathbf{P}_{NL}(\mathbf{E}) = \chi^{(2)}\mathbf{E}^2 + \chi^{(3)}\mathbf{E}^3 + \cdots = \mathbf{P}^{(2)} + \mathbf{P}^{(3)} + \cdots. \tag{2.65}$$

In this section, we will only consider terms up to third order in the electric field.

In the following treatment, we assume that the material is local in space and time. This implies that the polarization at one point of the material is not affected by the field at any other point. These restrictions can later be relaxed, but the phenomena that result will be the same.

If the sample is exposed to external fields with frequencies of ω_1 and ω_2, the total electric field in scalar form (for simplicity) is given by,

$$E = \frac{E^{\omega_1}}{2}e^{-i\omega_1 t} + \frac{E^{\omega_2}}{2}e^{-i\omega_2 t} + c.c., \tag{2.66}$$

and the polarization P, is given by

$$P = \frac{P^{2\omega_1}}{2}e^{-2i\omega_1 t} + \frac{P^{2\omega_2}}{2}e^{-2i\omega_2 t} + \frac{P^{\omega_1+\omega_2}}{2}e^{-i(\omega_1+\omega_2)t} + \cdots + c.c. \tag{2.67}$$

We take

$$P_{NL} = \chi^{(2)}E^2, \tag{2.68}$$

and in Equation 2.64 project out the Fourier component of interest. First let's calculate E^2 using Equation 2.66

$$\begin{aligned} E^2 &= \frac{(E^{\omega_1})^2}{4}e^{-2i\omega_1 t} + \frac{(E^{\omega_2})^2}{4}e^{-2i\omega_2 t} + \frac{E^{\omega_1}E^{-\omega_1}}{2} + \\ &\quad \frac{E^{\omega_1}E^{-\omega_2}}{4}e^{-i(\omega_1-\omega_2)t} + \cdots + c.c. \end{aligned} \tag{2.69}$$

To project out the $2\omega_1$ component we calculate the integral

$$\int_{-\infty}^{\infty} e^{-2i\omega_1 t}\left(P = \chi^{(2)}E^2\right)dt, \tag{2.70}$$

which yields

$$P^{2\omega_1} = \frac{1}{2}\chi^{(2)}\left(E^{\omega_1}\right)^2. \tag{2.71}$$

The coefficient of 1/2 in Equation 2.71 arises from squaring the field. This coefficient for other process is calculated in the same way – that is, the desired fourier component of the polarization is calculated after taking the superposition of input fields to the nth power for an nth-order response.

For the next case, P^0,

$$P^0 = \chi^{(2)}E^{-\omega_1}E^{\omega_1}. \tag{2.72}$$

Despite the fact that there was no applied static field, the driving field produces this field through nonlinear interaction with the sample. The same procedure applied to output frequency $\omega_1 + \omega_2$ gives

$$P^{\omega_1+\omega_2} = \chi^{(2)}E^{\omega_1}E^{\omega_2}. \tag{2.73}$$

Now, let's assume there are two fields E^ω and E^0. The field with zero frequency is treated separately because it is usually expressed as a real quantity. In this case, the total field is

$$E = \left(\frac{E^\omega}{2}e^{-i\omega t} + c.c.\right) + E^0. \tag{2.74}$$

Then, squaring Equation 2.74 leads to

$$E^2 = E^\omega E^0 e^{-i\omega t} + \frac{(E^\omega)^2}{4}e^{-2i\omega t} + \cdots, \tag{2.75}$$

whence

$$P^{(2)} = 2\chi^{(2)}E^\omega E^0. \tag{2.76}$$

$\chi^{(3)}$, in general, is written in terms of ω_1, ω_2 and ω_3 as the frequencies of the three incident fields. However, this general case is too complicated to serve as an example. So, for the sake of simplicity, let's assume there is only one field of frequency ω,

$$E = \frac{E^\omega}{2}e^{-i\omega t} + c.c. \tag{2.77}$$

From Equation 2.77,

$$E^3 = \frac{(E^\omega)^3}{8} e^{-3i\omega t} + \frac{3}{8}(E^\omega)^2 E^{-\omega} + \cdots + c.c., \tag{2.78}$$

whence,

$$P^{3\omega} = \frac{1}{4}(E^\omega)^3 \chi^{(3)}. \tag{2.79}$$

Following the same procedure for P^ω yields

$$P^\omega = \frac{3}{4}\chi^{(3)} E^\omega E^{-\omega} E^\omega, \tag{2.80}$$

where $E^\omega E^{-\omega} = |E^\omega|^2$. (Recall that $E^{\omega*} = E^{-\omega}$.)

To understand how nonlinear interactions lead to changes in the effective refractive index, we solve a specific problem in which a field with frequency ω interacts with a static field. Consider a beam that travels through a material between capacitor plates, which are used to apply a static electric field. Ignoring second-order nonlinearities (and higher than third order), we get

$$\begin{aligned}\mathbf{D}^\omega &= \mathbf{E}^\omega + 4\pi\chi^{(1)}\mathbf{E}^\omega + 8\pi\chi^{(2)} E^0 \mathbf{E}^\omega \\ &= \left(1 + 4\pi\chi^{(1)} + 8\pi\chi^{(2)} E^0\right)\mathbf{E}^\omega \\ &= \epsilon \mathbf{E}^\omega,\end{aligned} \tag{2.81}$$

where $\epsilon_0 = 1 + 4\pi\chi^{(1)}$ and

$$\epsilon = \epsilon_0 + 8\pi\chi^{(2)} E^0. \tag{2.82}$$

This indicates that the resulting field, generated by the interaction of incident field and the sample oscillates at frequency ω. The relationship between refractive index n and permittivity ϵ is given by

$$\begin{aligned}n &= \sqrt{\epsilon} \\ &= \epsilon_0^{1/2}\sqrt{1 + \frac{8\pi\chi^{(2)} E^0}{\epsilon_0}},\end{aligned} \tag{2.83}$$

which for small fields yields

$$n = n_0\left(1 + \frac{4\pi\chi^{(2)}}{\epsilon_0} E^0\right). \tag{2.84}$$

Finally,

$$n = n_0 + \frac{4\pi\chi^{(2)}}{n_0} E^0, \tag{2.85}$$

where $n_0 = \epsilon_0^2$. Equation 2.85 shows how the refractive index depends on the static field. The coefficient of E^0 in Equation 2.85 is called the electro-optic coefficient, n_1,

$$n_1 = \frac{4\pi \chi^{(2)}}{n_0}. \tag{2.86}$$

Considering now only the third order susceptibility, $\chi^{(3)}$, for only one applied field,

$$P^\omega = \chi^{(1)} E^\omega + \frac{3}{4} \chi^{(3)} |E^\omega|^2 E^\omega. \tag{2.87}$$

Once again we use Equation 2.56 in scalar form. ϵ is then given by

$$\epsilon = 1 + 4\pi \chi^{(1)} + \frac{3}{4}(4\pi)\chi^{(3)} |E^\omega|^2, \tag{2.88}$$

where we have neglected $\chi^{(2)}$, a good approximation for a centrosymmetric material such as a liquid. In analogy to Equation 2.83, the refractive index for the third-order response is

$$n = n_0 + \frac{3\pi \chi^{(3)}}{2n_0} |E^\omega|^2. \tag{2.89}$$

The optical kerr effect coefficient, n_2, is given by

$$n_2 = \frac{3\pi \chi^{(3)}}{2n_0}. \tag{2.90}$$

In general, depending on the symmetry properties of the material, either Equations 2.85 or 2.89 dominate. For example, in a centrosymmetric sample, where $\chi^{(2)}$ is zero, the refractive index is calculated via Equation 2.89.

2.3 Response Functions

In general, the polarization of a material depends on a response function which characterizes the material. For the first order case the polarization is

$$P_i^{(1)}(t) = \int_{-\infty}^{\infty} T_{ij}^{(1)}(t;\tau) E_j(\tau) d\tau, \tag{2.91}$$

where we have assumed that the response is spatially local, and the response function is T_{ij}.[1] This form for the polarization takes into account how fields

---

[1] In this section limits of integrals are assumed to be from $-\infty$ to ∞ unless otherwise stated.

Lecture Notes in Nonlinear Optics 37

at different points in time, τ, will contribute to the material's response at time t. To understand some of the properties of response functions, we will begin with the first order case, and then we will generalize the approach to higher orders.

2.3.1 Time Invariance

One of the most important axioms of physics is that the laws of physics are invariant under time translation. So if we let $t \to t+t_0$ in, $P_i^{(1)}$ and E_j the laws of physics should remain unchanged. Advancing the polarization by $t \to t+t_0$ in equation 2.91 yields

$$P_i^{(1)}(t+t_0) = \int T_{ij}^{(1)}(t+t_0;\tau)E_j(\tau)d\tau. \tag{2.92}$$

Advancing the polarization and fields by $t \to t+t_0$ yields:

$$P_i^{(1)}(t+t_0) = \int T_{ij}^{(1)}(t;\tau)E_j(\tau+t_0)d\tau. \tag{2.93}$$

Making the substitution $\tau \to \tau'$ in equation 2.92 and $\tau' = \tau + t_0$ in equation 2.93, equations 2.92 and 2.93

$$\int T_{ij}^{(1)}(t+t_0;\tau')E_j(\tau')d\tau' = \int T_{ij}^{(1)}(t;\tau'-t_0)E_j(\tau')d\tau'. \tag{2.94}$$

If equation 2.94 is to hold for all $E_j(\tau')$, then it follows that

$$T_{ij}^{(1)}(t+t_0;\tau') = T_{ij}^{(1)}(t;\tau'-t_0). \tag{2.95}$$

with $t = 0$ and $t_0 \to t'$ we get,

$$T_{ij}^{(1)}(t';\tau') = T_{ij}^{(1)}(0;\tau'-t'), \tag{2.96}$$

we see that the response depends only on $\tau' - t'$. Now changing from $T(t,\tau)$ (a function of two variables) to $R(\tau-t)$ (a function of one variable) we can rewrite the polarization as given by equation 2.91

$$P_i^{(1)}(t) = \int R_{ij}^{(1)}(t-\tau)E_j(\tau)d\tau. \tag{2.97}$$

If we enforce causality by setting $R_{ij}^{(1)}(t-\tau) = 0$ for $\tau > t$ (i.e. demanding that a polarization is possible only after the application of a field and defining $\tau' = t - \tau$, we get

$$P_i^{(1)}(t) = \int_0^\infty R_{ij}^{(1)}(\tau')E_j(t-\tau')d\tau'. \tag{2.98}$$

The result is a description of the polarization of a material at time t due to fields acting locally over all times in the past.

2.3.2 Fourier Transforms of Response Functions: Electric Susceptibilities

In the previous section we derived the polarization of a material as a function of time due to an electric field. Often it is more convenient and appropriate to consider the polarization in frequency space. The Fourier Transforms of the electric field are defined by

$$E_j(t) = \int d\omega \tilde{E}_j(\omega)e^{-i\omega t} \tag{2.99}$$

and

$$\tilde{E}_j(\omega) = \frac{1}{2\pi}\int dt E_j(t)e^{i\omega t}. \tag{2.100}$$

Substituting $E_j(t)$ into equation 2.98 yields

$$P_i^{(1)}(t) = \int R_{ij}^{(1)}(\tau')\int \tilde{E}_j(\omega)e^{-i\omega(t-\tau')}d\omega d\tau', \tag{2.101}$$

which we can regroup into the form,

$$P_i^{(1)}(t) = \int \left(\int R_{ij}^{(1)}(\tau')e^{-i\omega\tau'}d\tau'\right)\tilde{E}_j(\omega)e^{-i\omega t}d\omega. \tag{2.102}$$

The term in parentheses in equation 2.102 is the inverse Fourier transform of $R_{ij}^{(1)}$, which we define as

$$\chi_{ij}^{(1)}(\omega) = \int R_{ij}^{(1)}(\tau')e^{-i\omega\tau'}d\tau', \tag{2.103}$$

the first order electric susceptibility. Substituting equation 2.103 into equation 2.102, we find

$$P_i^{(1)}(t) = \int \chi_{ij}^{(1)}(\omega)\tilde{E}_j(\omega)e^{-i\omega t}d\omega. \tag{2.104}$$

Next, we transform the polarization into frequency space according to,

$$P_i^{(1)}(\omega_\sigma) = \frac{1}{2\pi}\int P_i^{(1)}(t)e^{i\omega_\sigma t}dt \tag{2.105}$$

Lecture Notes in Nonlinear Optics

substituting equation 2.104 into equation 2.105 yields

$$P_i^{(1)}(\omega_\sigma) = \frac{1}{2\pi} \int \int \chi_{ij}^{(1)}(\omega)\tilde{E}_j(\omega)e^{-i\omega t}e^{i\omega_\sigma t}d\omega dt. \tag{2.106}$$

Recall that the dirac delta function can be expressed as,

$$\delta(\omega - \omega_\sigma) = \frac{1}{2\pi} \int e^{-it(\omega - \omega_\sigma)}dt, \tag{2.107}$$

we can rewrite equation 2.106 as

$$P_i^{(1)}(\omega_\sigma) = \int \chi_{ij}^{(1)}(\omega)\tilde{E}_j(\omega)\delta(\omega - \omega_\sigma)d\omega. \tag{2.108}$$

Performing the integration yields

$$P_i^{(1)}(\omega_\sigma) = \chi_{ij}^{(1)}(\omega_\sigma)\tilde{E}_j(\omega_\sigma). \tag{2.109}$$

Thus we have found that in the frequency representation, the polarization is a simple relationship between the electric field and the electric susceptibility, where the susceptibility is defined as the Fourier transform of the response function.

2.3.3 A Note on Notation and Energy Conservation

Throughout our book we will be focusing on electric susceptibilities, which in general can depend on multiple frequencies, and the order in which they appear. In anticipation of future needs we introduce the notation $\chi^{(n)}(-\omega_{out};\omega_{in})$.

The angular frequency to the left of the semi colon represents the outgoing wave, or the frequency we are measuring in the experiment, the angular frequencies on the right represent all the incoming fields. The negative sign on the outgoing field represents energy leaving the material, where as the positive frequencies to the right of the semicolon represent energy entering the system. This notation explicitly denotes energy conservation where $\sum \omega_{in} + \sum \omega_{out} = 0$.

For the first order susceptibility the polarization is expressed as

$$P_i^{(1)}(\omega) = \chi_{ij}^{(1)}(-\omega;\omega)\tilde{E}_j(\omega), \tag{2.110}$$

where the sum of input frequencies and output frequencies is simply $-\omega + \omega = 0$.

2.3.4 Second Order Polarization and Susceptibility

For the second order case, we begin with the relationship between the polarization $P^{(2)}(t)$ and response function $R^{(2)}_{ijk}$,

$$P^{(2)}_i(t) = \int\int R^{(2)}_{ijk}(\tau_1,\tau_2) E_j(t-\tau_1) E_k(t-\tau_2) d\tau_1 d\tau_2, \qquad (2.111)$$

where $R^{(2)}_{ijk}(\tau_1,\tau_2) = 0$ if τ_1 or $\tau_2 = 0$. Substituting the Fourier transform again of the electric field using equation 2.109, we get

$$\begin{aligned}
P^{(2)}_i(t) &= \int\int R^{(2)}_{ijk}(\tau_1,\tau_2)\left(\int \tilde{E}_j(\omega_1)e^{-i\omega_1(t-\tau_1)}d\omega_1\right) \\
&\quad \times \left(\int \tilde{E}_k(\omega_2)e^{-i\omega_2(t-\tau_2)}d\omega_2\right) d\tau_1 d\tau_2 \\
&= \int\int\int\int R^{(2)}_{ijk}(\tau_1,\tau_2)\tilde{E}_j(\omega_1)\tilde{E}_k(\omega_2) \\
&\quad \times e^{-i\omega_1(t-\tau_1)}e^{-i\omega_2(t-\tau_2)}d\omega_1 d\omega_2 d\tau_1 d\tau_2 \\
&= \int\int\int\int d\omega_1 d\omega_2 d\tau_1 d\tau_2 R^{(2)}_{ijk}(\tau_1,\tau_2)e^{-i\omega_1\tau_1}e^{-i\omega_2\tau_2} \\
&\quad \times \tilde{E}_j(\omega_1)\tilde{E}_k(\omega_2)e^{-i\omega_1 t}e^{-i\omega_2 t}.
\end{aligned} \qquad (2.112)$$

In a manner similar to the derivation of equation 2.109 we can simplify the expression by introducing the second order susceptibility,

$$\chi^{(2)}_{ijk}(\omega_1,\omega_2) = \int\int d\tau_1 d\tau_2 R^{(2)}_{ijk}(\tau_1,\tau_2) e^{-i\omega_1\tau_1} e^{-i\omega_2\tau_2}. \qquad (2.113)$$

The second-order polarization, then, is related to the second-order susceptibility,

$$P^{(2)}_i(t) = \int\int d\omega_1 d\omega_2 \chi^{(2)}_{ijk}(\omega_1,\omega_2)\tilde{E}_j(\omega_1)\tilde{E}_k(\omega_2) e^{-i\omega_1 t}e^{-i\omega_2 t}. \qquad (2.114)$$

We once again take the inverse Fourier transform of $P^{(2)}_i(t)$, yielding

$$P^{(2)}_i(\omega) = \frac{1}{2\pi}\int P^{(2)}_i(t) e^{i\omega t} dt. \qquad (2.115)$$

substituting equation 2.114 into equation 2.115 yields,

$$P_i^{(2)}(\omega) = \int\int\int d\omega_1 d\omega_2 dt \chi_{ijk}^{(2)}(\omega_1,\omega_2)\tilde{E}_j(\omega_1)\tilde{E}_k(\omega_2)e^{i(\omega_1+\omega_2-\omega)t} \quad (2.116)$$

$$= \int\int d\omega_1 d\omega_2 \chi_{ijk}^{(2)}(\omega_1,\omega_2)\tilde{E}_j(\omega_1)\tilde{E}_k(\omega_2)\delta(\omega_1+\omega_2-\omega). \quad (2.117)$$

Performing the integration we get,

$$P_i^{(2)}(\omega) = \chi_{ijk}^{(2)}(-\omega;\omega_1,\omega_2)\tilde{E}_j(\omega_1)\tilde{E}_k(\omega_2), \quad (2.118)$$

where again, we use the notation where all frequency arguments sum to zero.

2.3.5 n^{th} Order Polarization and Susceptibility

We can easily generalize the method from the previous section to find higher-order polarizations and susceptibilities. The general form of the nth-order polarization is

$$P_i^{(n)}(\omega) = \chi_{ijkl...}^{(n)}(-\omega;\omega_1,\omega_2,....\omega_n)E_j(\omega_1)E_k(\omega_2)... \quad (2.119)$$

where $\chi^{(n)}$ is given by

$$\chi_{ijk...}^{(n)}(-\omega;\omega_1,\omega_2,....\omega_n) = \int d\tau_1 \int d\tau_2 .... \int d\tau_n R_{ijk...}^{(n)}(\tau_1,\tau_2.....\tau_n)$$
$$\times\ e^{i\omega_1\tau_1}e^{i\omega_2\tau_2}....e^{i\omega_n\tau_n}. \quad (2.120)$$

2.3.6 Properties of Response Functions

Now that we have found a general relationship between the polarization and the applied fields, we now consider the properties of response functions. Response functions have full permutation symmetry, meaning that if we have a polarization

$$P_i^{(n)}(t) = \int d\tau_1 \int d\tau_2 .. R_{ijk...}^{(n)}(t;\tau_1,\tau_2,....\tau_n)E_j(\tau_1)E_k(\tau_2)..... \quad (2.121)$$

we can exchange two spatial and time components simultaneously and not affect the polarization as follows, Exchanging τ_1 and τ_2, and also exchanging j and k gives

$$P_i^{(n)}(t) = \int d\tau_1 \int d\tau_2 .. R_{ikj...}^{(n)}(t;\tau_2,\tau_1,....\tau_n)E_k(\tau_2)E_j(\tau_1)..... \quad (2.122)$$

Note that $P_i^{(n)}(t)$ in both equations 2.121 and 2.122 must be the same because τ_1, τ_2, j and k are dummy indicies and so can be renamed in this war without affecting the integrals and sums. But, since $E_j(\tau_1)E_k(\tau_2) = E_k(\tau_2)E_j(\tau_1)$, then $R_{ijk...}^{(n)}(t; \tau_1, \tau_2...) = R_{ikj...}^{(n)}(t; \tau_2, \tau_1...)$. Thus, the response function is unchanged when $k \leftrightarrow j$ and $\tau_1 \leftrightarrow \tau_2$. This is called permutation symmetry. Exchange of any two indicies(excluding the first one) and the corresponding two time arguements leaves the response function unchanged. As an example, consider the 2rd order response function,

$$R_{xyx}^{(2)} = R_{xxy}^{(2)} \neq R_{yxx}^{(2)}, \tag{2.123}$$

where the third permutation is not equivalent because the first index is involved.

As we saw in section 2.2 the definitions of the fields leads to numerical prefactors in the relationship between $P_\mu^{(n)}(\omega)$ and the fields. In the most general case for the notation used here, the polarization is given by

$$\begin{aligned}P_\mu^{(n)}(\omega) &= \sum_n \sum_{\alpha_1,\alpha_2...} K(-\omega; \omega_1, \omega_2, ...\omega_n) \chi_{\mu,\alpha_1,\alpha_2....}^{(n)}(-\omega; \omega_1, \omega_2, ...\omega_n) \\ &\times E_{\alpha_1}(\omega_1) E_{\alpha_2}(\omega_2)...E_{\alpha_n}(\omega_n)\end{aligned} \tag{2.124}$$

where $K(-\omega; \omega_1, \omega_2....\omega_n)$ is defined in our convention to be

$$K(-\omega; \omega_1, \omega_2....\omega_n) = 2^{l+m-n} \Pi. \tag{2.125}$$

Π is the distinct number of permutations of $\omega_1, \omega_2...\omega_n$, n is the order of the nonlinearity, m is the number of times $\omega_i = 0$, and $l = 1$ if $\omega \neq 0$ and $l = 0$ if $\omega = 0$.

Example: Optical Kerr Effect

In the Optical Kerr Effect we have two photons of frequency ω incident on a sample and two emitted photons as diagramed by Fig 2.10. This diagram corresponds to a third order susceptibility $\chi^{(3)}(-\omega; \omega, -\omega, \omega)$. There are no zero frequencies so $m = 0$. Since the output is nonzero, $l = 1$. The order of the nonlinearity is 3 and the number of distinct permutations of the frequencies $(\omega, -\omega, \omega)$ is $\Pi = 3$. Therefore

$$K = 2^{1+0-3}(3) = \frac{3}{4}, \tag{2.126}$$

which gives

$$P^{(3)}(\omega) = \frac{3}{4} \chi^{(3)}(-\omega; \omega, -\omega, \omega) E(\omega) E(-\omega) E(\omega), \tag{2.127}$$

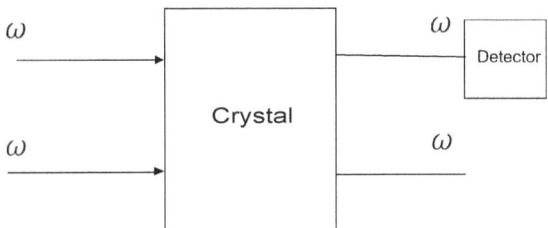

Figure 2.10: In the optical Kerr effect two beams of ω enter the sample and one causes a change in the index of refraction of the other, which we measure on the other side.

the result we got in equation 2.80

Problem 2.3-1(a): An electric field pulse of the form $E(t) = t_0 E_0 \delta(t)$ is applied to a material. δ is the Dirac Delta function, t_0 has units of time and E_0 has units of electric field strength. The polarization responds as a damped oscillator of the form $P^{(1)}(t) = P_0 e^{-t/\tau} \sin \omega_0 t$, where ω_0 is the frequency of oscillation and τ the damping time. Use this result to determine the response function $R^{(1)}(t)$ and from this determine the linear susceptibility $\chi^{(1)}(-\omega;\omega)$. Compare the result you get with $\chi^{(1)}(-\omega;\omega)$ for the nonlinear spring.

(b): For a response function of the form

$$R^{(2)}(t_1,t_2) = R_0 \theta(t_1)\theta(t_2) e^{-t_1/\tau} \sin(\omega_0 t_1) e^{-t_2/\tau} \sin(\omega_0 t_2), \quad (2.128)$$

calculate $\chi^{(2)}(\omega_1,\omega_2)$ AND $P^{(2)}(t)$. To calculate $P^{(2)}(t)$, assume that the electric field is a step function of the form $E(t) = E_0 \theta(t)$.

(c): Consider a nonlinear spring with only $k^{(1)} \neq 0$ and $k^{(2)} \neq 0$. If $\Gamma = 0$ and there is no driving field, show that

$$x = A e^{-i\omega_0 t} + \delta e^{-i2\omega_0 t}$$

is a solution of the equations of motion when $\delta \ll A$. Show how δ depends on A, $k^{(1)}$ and $k^{(2)}$. Do the results make physical sense in light of what you got in parts (a) and (b)?

2.4 Properties of the Response Function

The nonlinear polarization $P(\omega)$ can be described in the frequency domain, as in the case when combinations of monochromatic fields are applied to an optical material. The optical nonlinear polarization can also be written in the time domain as $P(t)$. This is especially effective when considering problems where the applied fields appear as short pulses. To get a better understanding of the time domain polarization let's look at the case of a material with a linear response.

In the case of linear polarization we had,

$$P^{(1)}(t) = \int_0^\infty R^{(1)}(\tau) E(t-\tau) d\tau. \tag{2.129}$$

Here $P^{(1)}(t)$ and $E(t-\tau)$ are observed quantities and are therefore real, which implies that $R^{(1)}(\tau)$ is also real.

The linear susceptibility, $\chi^{(1)}(\omega)$, is a complex quantity with real and imaginary parts, and is given by,

$$\chi^{(1)}(\omega) = \int_0^\infty R^{(1)}(\tau) e^{i\omega t} d\tau. \tag{2.130}$$

Here $R^{(1)}(\tau)$ is the linear response function that relates the polarization at a time t to an applied electric field at some earlier time $t-\tau$. The lower limit of integration is set to zero rather than $-\infty$ because we require that $P^{(1)}(t)$ does not depend on the future values of $E(t)$; only the past values of $E(t)$ can affect $P^{(1)}(t)$. This is equivalent to setting $R^{(1)}(\tau) = 0$ for $\tau < 0$ to impose the "causality condition." From equation 2.130 and the fact that $R^{(1)}(\tau)$ is real we get the following relationship

$$\chi^{(1)}(\omega_i)^* = \chi^{(1)}(-\omega_i). \tag{2.131}$$

2.4.1 Kramers-Kronig

To develop the Kramers-Kronig equations we will evaluate the integral

$$\int_{-\infty}^\infty \frac{\chi^{(1)}(\omega')}{\omega' - \omega} d\omega'. \tag{2.132}$$

We will treat ω as complex for the purpose of this integration, with $\omega \to \omega + i\omega_I$. To evaluate this integral, we consider the contour integral shown in Figure 2.11. The contribution to the contour integral from the large semicircular path goes to zero as $R \equiv |\omega| \to \infty$ since the denominator goes to ∞ as $R \to \infty$.

Lecture Notes in Nonlinear Optics 45

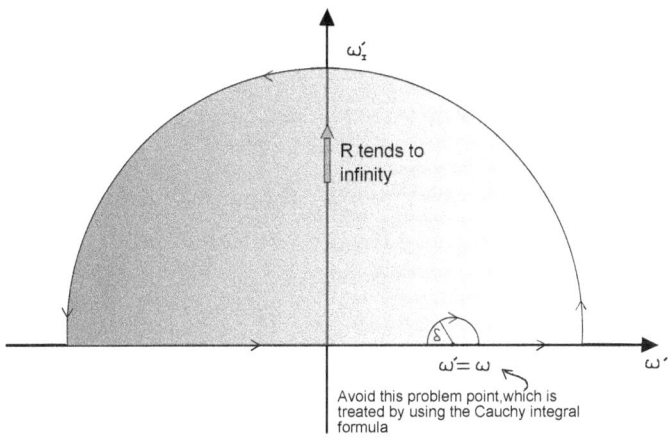

Figure 2.11: The value of the contour integral is composed of contributions from the three major pieces of the path: (1) The semicircle of increasing radius for which $\omega_I > 0$. (2) The real axis, which avoids the singular point $\omega = \omega'$ by detouring around on a semicircular arc. (3) The pole $\omega = \omega'$ is integrated using Cauchy's integral formula.

The principal value of an integral whose path passes through a singularity is defined as the part of the integral that excludes the singular point $\omega = \omega'$, and takes the form

$$\lim_{\delta \to 0}\left(\int_{-\infty}^{\omega'-\delta}\frac{\chi^{(1)}(\omega')}{\omega'-\omega}d\omega' + \int_{\omega'+\delta}^{\infty}\frac{\chi^{(1)}(\omega')}{\omega'-\omega}d\omega'\right). \qquad (2.133)$$

The path of integration, shown in Figure 2.11, is drawn in a way that all poles are excluded. Thus, Cauchy's Integral Formula yields

$$\oint_{-\infty}^{\infty}\frac{\chi^{(1)}(\omega')}{\omega'-\omega}d\omega' = 0. \qquad (2.134)$$

This contour integral can be separated in the three paths as shown in Figure 2.11: (1) the integral along the real axis that excludes the singularity (i.e. the principle value integral); (2) the small semicircular path that circles the pole; and (3) the large semicircular path at infinity. According to Equation 2.134 These three terms give,

$$\int_{-\infty}^{\infty}\frac{\chi^{(1)}(\omega')}{\omega'-\omega}d\omega' + i\pi\chi^{(1)}(\omega) + \int_{R}\frac{\chi^{(1)}(\omega')}{\omega'-\omega}d\omega' = 0. \qquad (2.135)$$

The last term in Equation 2.135 vanishes as R tends to ∞. Therefore, Equation 2.135 yeilds,

$$\chi^{(1)}(\omega) = \frac{-i}{\pi} \int_{-\infty}^{\infty} \frac{\chi^{(1)}(\omega')}{\omega' - \omega} d\omega'. \tag{2.136}$$

Since we can write $\chi^{(1)} = \chi_R^{(1)} + \chi_I^{(1)}$; then, the fact that $-i(\chi_R^{(1)} + \chi_I^{(1)}) = -i\chi_R^{(1)} + \chi_I^{(1)}$ leads to the Kramers-Kronig relationships,

$$\text{Re}[\chi^{(1)}(\omega)] = \frac{1}{\pi} \int_{-\infty}^{\infty} \text{Im}\left[\frac{\chi^{(1)}(\omega')}{\omega' - \omega}\right] d\omega', \tag{2.137}$$

and

$$\text{Im}[\chi^{(1)}(\omega)] = \frac{-1}{\pi} \int_{-\infty}^{\infty} \text{Re}\left[\frac{\chi^{(1)}(\omega')}{\omega' - \omega}\right] d\omega'. \tag{2.138}$$

The Kramers-Kronig equations relate the real part of the nonlinear susceptibility to the imaginary part provided that the imaginary part is known over a broad enough range of angular frequencies, ω, to perform the required integration. Similarly, the imaginary part of $\chi^{(1)}$ depends on integral of the real part. The imaginary part of χ is dispersive and since it is more common to measure absorption spectra, Equation 2.137 can be used to predict the frequency dependence of the refractive index.

The Kramers-Kronig equations can be re-expressed in a more useful form, as follows. From equation 2.131, we find that

$$\text{Re}[\chi^{(1)}(-\omega)] = \text{Re}[\chi^{(1)}(\omega)], \tag{2.139}$$
$$\text{Im}[\chi^{(1)}(-\omega)] = -\text{Im}[\chi^{(1)}(\omega)], \tag{2.140}$$

so, we can rewrite Equation 2.138 as

$$\text{Im}[\chi^{(1)}(\omega)] = -\frac{1}{\pi} \int_{-\infty}^{0} \frac{\text{Re}[\chi^{(1)}(\omega')]}{\omega' - \omega} d\omega' - \frac{1}{\pi} \int_{0}^{\infty} \frac{\text{Re}[\chi^{(1)}(\omega')]}{\omega' - \omega} d\omega', \tag{2.141}$$

which is equivalent to

$$\text{Im}[\chi^{(1)}(\omega)] = \frac{1}{\pi} \int_{0}^{\infty} \frac{\text{Re}[\chi^{(1)}(\omega')]}{\omega' + \omega} d\omega' - \frac{1}{\pi} \int_{0}^{\infty} \frac{\text{Re}[\chi^{(1)}(\omega')]}{\omega' - \omega} d\omega'. \tag{2.142}$$

This gives the final form

$$\text{Im}[\chi^{(1)}(\omega)] = -\frac{2\omega}{\pi} \int_{0}^{\infty} \frac{\text{Re}[\chi^{(1)}(\omega')]}{\omega'^2 - \omega^2} d\omega'. \tag{2.143}$$

The other complex component can be found by the following method,

$$\text{Re}[\chi^{(1)}(\omega)] = \frac{1}{\pi}\int_{-\infty}^{0} \text{Im}\left[\frac{\chi^{(1)}(\omega')}{\omega'-\omega}\right]d\omega' + \frac{1}{\pi}\int_{0}^{\infty} \text{Im}\left[\frac{\chi^{(1)}(\omega')}{\omega'-\omega}\right]d\omega' \quad (2.144)$$

$$\text{Re}[\chi^{(1)}(\omega)] = \frac{1}{\pi}\int_{-\infty}^{0} \text{Im}\left[\frac{\chi^{(1)}(-\omega')}{-\omega'-\omega}\right](-d\omega') + \frac{1}{\pi}\int_{0}^{\infty} \text{Im}\left[\frac{\chi^{(1)}(\omega')}{\omega'-\omega}\right]d\omega' \quad (2.145)$$

$$\text{Re}[\chi^{(1)}(\omega)] = -\frac{1}{\pi}\int_{0}^{\infty} \text{Im}\left[\frac{\chi^{(1)}(\omega')}{\omega'+\omega}\right](d\omega') + \frac{1}{\pi}\int_{0}^{\infty} \text{Im}\left[\frac{\chi^{(1)}(\omega')}{\omega'-\omega}\right]d\omega' \quad (2.146)$$

$$\text{Re}[\chi^{(1)}(\omega)] = \frac{2\omega}{\pi}\int_{0}^{\infty} \text{Im}\left[\frac{\chi^{(1)}(\omega')}{\omega'^2-\omega^2}\right]d\omega'. \quad (2.147)$$

If we look at the Kramers-Kronig relations using a viewpoint of Hilbert transform filters of the kind,

$$H(\omega) = \int_{-\infty}^{\infty} e^{-i\omega t} h(t)dt = \int_{-\infty}^{\infty} (h(t)\cos(\omega t) - ih(t)\sin(\omega t))dt, \quad (2.148)$$

the concept of causality is captured in the statement of the Kramers-Kronig relationships. Applying a frequency domain convolution amounts to sliding a kernel function, in this case projecting out the even and odd parts of the causal response of $\chi^{(1)}(\omega)$ which correspond to the real and imaginary pieces of the frequency response. For example,

$$\int_{-\infty}^{\infty} \chi^{(1)}(\omega-\omega')H(\omega)d\omega = \int_{-\infty}^{\infty} \chi^{(1)}(\omega-\omega')[\cos(\omega t) + i\sin(\omega t)]d\omega. \quad (2.149)$$

The Kramers Kroning relationships can help determine when a response is causal or not realistic in time. We can tell whether the response is causal by testing whether the real and imaginary parts of the Hilbert transforms are equal to each other.

2.4.2 Permutation Symmetry

In general for different orders of nonlinearity, χ will be a tensor of many indices (3 in the case of 2nd order nonlinear susceptibility) and will depend on combinations of the input and output frequencies. The negatives of these frequencies (representing outgoing fields) add additional members to the set possible of possible frequency combinations contributing to each tensor element. Fortunately, we do not need to keep track of all these separate members by using symmetry properties.

From the defining relationship expressing the polarization as a function of space and time, the time dependence of the sum-frequency polarization for two incident monochromatic waves of frequencies ω_1 and ω_2 is given by

$$P(r,t) = P_i(\omega_1 + \omega_2)e^{-i(\omega_1+\omega_2)t} + P_i(-\omega_1 - \omega_2)e^{i(\omega_1+\omega_2)t}. \tag{2.150}$$

The function $P(r,t)$ is real since it is a physical and therefore (theoretically) measurable quantity and by equation 2.150 we get

$$P_i(-\omega_1 - \omega_2) = P_i(\omega_1 + \omega_2)^*. \tag{2.151}$$

The electric fields are also real, leading to

$$\begin{aligned} E_j(-\omega_1) &= E_j(\omega_1)^*, \\ E_k(-\omega_2) &= E_k(\omega_2)^*. \end{aligned} \tag{2.152}$$

From Equations 2.151 and 2.152, we get

$$P_i(-\omega_1 - \omega_2) = \chi^{(2)}_{ijk}(-\omega_1 - \omega_2, -\omega_1, -\omega_2)E_j(-\omega_1)E_k(-\omega_2), \tag{2.153}$$

and

$$P_i(\omega_1 + \omega_2)^* = \chi^{(2)}_{ijk}(\omega_1 + \omega_2, \omega_1, \omega_2)^* E_j(\omega_1)^* E_k(\omega_2)^*, \tag{2.154}$$

with an implied sum over i and j. We can conclude that the positive and negative components of the nonlinear susceptibility tensor have the symmetry

$$\chi^{(2)}_{ijk}(-\omega_1 - \omega_2, -\omega_1, -\omega_2) = \chi^{(2)}_{ijk}(\omega_1 + \omega_2, \omega_1, \omega_2)^*, \tag{2.155}$$

which reduces the number of independent components needed to describe the susceptibility tensor χ.

Full Permutation Symmetry

Some of the indices of the susceptibility tensor can be exchanged without affecting the polarizability. For example, the quantity $\chi_{ijk}E_jE_k$ will lead to the same polarization response when the two fields are interchanged, i.e. when $\chi_{ijk}E_jE_k \rightarrow \chi_{ijk}E_kE_j = \chi_{ikj}E_jE_k$ because j and k are dummy indices and can be renamed. Thus, $\chi_{ijk} = \chi_{ikj}$. When the electric fields are at different frequencies ω_n and ω_m, the product of the fields is of the form $E_k(\omega_n)E_j(\omega_m)$, so the polarization remains unchanged when i and j are interchanged along with ω_n and ω_m, thus,

$$\chi^{(2)}_{ijk}(\omega_n + \omega_m, \omega_n, \omega_m) = \chi^{(2)}_{ikj}(\omega_n + \omega_m, \omega_m, \omega_n). \tag{2.156}$$

Lecture Notes in Nonlinear Optics 49

Equation 2.156 is called intrinsic permutation symmetry.

For lossless media, the components of χ are all guaranteed to be real and the interchange of any two indices (including the first one) along with the corresponding frequencies leaves the tensor unchanged. This is called full permutation symmetry and holds for non-dispersive materials.

Kleinman's Symmetry

When the Bohr frequencies in a material are much larger that the frequencies of all the photons in a nonlinear process, called the off-resonance regime, all of the photon frequencies can be approximated to be small enough to set them to zero frequency. In this case, the interchange of any two frequencies leaves $\chi^{(2)}_{ijk}$ unchanged, so equivalently, by the full intrinsic permutation symmetry condition, any two indices can be interchanged. Therefore, the full intrinsic permutation symmetry condition,

$$\chi^{(2)}_{ijk}(\omega_3 = \omega_1 + \omega_2) = \chi^{(2)}_{jki}(\omega_1 = -\omega_2 + \omega_3) = \chi^{(2)}_{kij}(\omega_2 = \omega_3 - \omega_1) =$$
$$\chi^{(2)}_{ikj}(\omega_3 = \omega_2 + \omega_1) = \chi^{(2)}_{kji}(\omega_2 = -\omega_1 + \omega_3) = \chi^{(2)}_{jik}(\omega_1 = \omega_3 - \omega_2)$$
(2.157)

becomes

$$\chi^{(2)}_{ijk} = \chi^{(2)}_{jki} = \chi^{(2)}_{kij} = \chi^{(2)}_{ikj} = \chi^{(2)}_{kji} = \chi^{(2)}_{jik}. \tag{2.158}$$

Contracted Notation

In order to simplify the notation when discussing systems with Kleinman Symmetry or second-harmonic generation (which is fully symmetric in its indices since ω_n and ω_m are equal), we can introduce the second harmonic tensor d_{ijk}, which is defined as

$$d_{ijk} = \frac{1}{2}\chi^{(2)}_{ijk}. \tag{2.159}$$

Given permutation symmetry, and the fact that the two incident photons are of the same frequency, the second two indices can be interchanged without changing the polarization. This allows us to define a 3 × 6 contracted matrix, where:

- For index (j,k)=(1,1) the second index is 1
- For index (j,k)=(2,2) the second index is 2

- For index (j,k)=(3,3) the second index is 3
- For the index (j,k)=(1,2) or if (j,k)=(2,1) the second index is 4
- For index (j,k)=(1,3) or (j,k)=(3,1) the second index is 5
- for index (j,k)=(1,2) or (j,k)=(2,1) the second index is 6

This gives the contracted tensor d_{il}, where i goes from 1 to 3, and l is between 1 and 6. Under Kleinman symmetry, the second harmonic tensor is given by,

$$d_{il} = \begin{bmatrix} d_{11} & d_{12} & d_{13} & d_{14} & d_{15} & d_{16} \\ d_{16} & d_{22} & d_{23} & d_{24} & d_{14} & d_{12} \\ d_{15} & d_{24} & d_{33} & d_{23} & d_{13} & d_{14} \end{bmatrix}.$$

Problem 2.4-1(a): Two light beams of frequency ω_1 and ω_2 impinge on a nonlinear material with a scalar and spatially local response ($\chi^{(n)}$ will not depend on $\vec{k}$). Calculate the Fourier amplitude $P^{(4)}_{2\omega_1-2\omega_2}$ in terms of $\chi^{(4)}$, E^{ω_1} and E^{ω_2}. In other words, write an expression of the form $P^{(4)}_{2\omega_1-2\omega_2} = K\chi^{(4)}(-(2\omega_1-2\omega_2);\ldots)E^{\omega_1}\ldots$

(b): Two high-intensity light beams of frequency ω_1 and ω_2 are launched into a centrosymmetric material. A spectrometer finds that light at frequency $\omega_1 - 2\omega_2$ is generated for an arbitrary set of input frequency ω_1 and ω_2 only when a static electric field is applied to the sample. Write an expression in a form as you did in (a) for the polarization at $\omega_1 - 2\omega_2$ for the lowest-order nonlinearity that could be responsible for the effect. Be sure to evaluate K. How must the intensity of the light at frequency $\omega_1 - 2\omega_2$ depend on the applied static voltage?

Problem 2.4-2(a): A physicist and an engineer set out to measure the $\chi^{(2)}$ and $\chi^{(3)}$ tensor of a material using a laser that is far off-resonance. The engineer runs to the lab and measures all 27 $\chi^{(2)}$ tensor components and all 81 $\chi^{(3)}$ tensor components. The physicist uses full permutation symmetry to eliminate redundant measurements. How many measurements are required to determined all tensor components? Make sure to list all of the tensor components and how they are related.

(b): If the material is isotropic, how many measurements are required in each case? Hint: apply the inversion operation.

Problem 2.4-3: The figures below show the ground state charge distributions of several molecules. In each case, use symmetry considerations to determine when either $\chi^{(2)}_{xxx}$ and/or $\chi^{(2)}_{yyy}$ are disallowed.

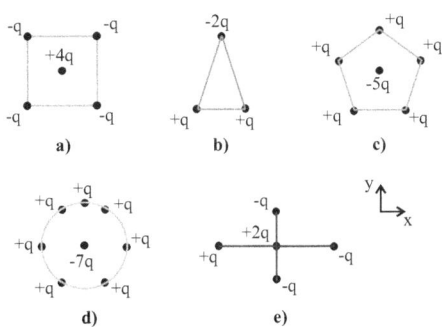

Chapter 3

Nonlinear Wave Equation

3.1 General Technique

In this section we develop a general technique to obtain the nonlinear wave equation. Subsequently, we will apply this technique to different nonlinear-optical processes. First we need to make a series of physically-reasonable assumptions to simplify the calculation, as follows.

We assume that there are no free charges present in space, $\rho_f = 0$; however, there are bound charges in any material. These bound charges will be the source of nonlinear polarization. Likewise, we assume that there are no free or induced currents, $\vec{J} = 0$. In fact, we do not consider any magnetic properties of the material, which implies $\vec{B} = \vec{H}$. This assumption will be valid in a broad range of materials; however, we must exercise care in cases where this assumption fails.

Secondly, recall from electrostatics, that the electric displacement field, $\vec{D}$, can be written as $\vec{D} = \vec{E} + 4\pi\vec{P}$. Since every material consists of atoms or molecules, applying an electric field results in an induced polarization in the material, which leads to generation of an internal field. To get the wave equation inside the material, we substitute the fields into the Maxwell's equations. The Maxwell's equations are,

$$
\begin{aligned}
\nabla \cdot \vec{D} &= 4\pi\rho_f, \\
\nabla \cdot \vec{B} &= 0, \\
\nabla \times \vec{E} &= -\frac{1}{c}\frac{\partial \vec{B}}{\partial t}, \\
\nabla \times \vec{H} &= \frac{4\pi}{c}\vec{J} + \frac{1}{c}\frac{\partial \vec{D}}{\partial t}.
\end{aligned}
\tag{3.1}
$$

Using the above approximations and assumptions, the last two Maxwell's equations yield,

$$\nabla \times \nabla \times \vec{E} + \frac{1}{c^2}\frac{\partial^2 \vec{D}}{\partial t^2} = 0. \tag{3.2}$$

The polarization of the material is of the form

$$P_i = \chi_{ij}^{(1)} E_j + P_i^{NL}. \tag{3.3}$$

We substitute the polarization into the electric displacement field and then into the Maxwell's wave equation (Equation 3.2). The linear term of the polarization yields the linear wave equation,

$$\nabla \times \nabla \times \vec{E} + \frac{\epsilon}{c^2}\frac{\partial^2 \vec{E}}{\partial t^2} = 0, \tag{3.4}$$

where $\epsilon = 1 + 4\pi\chi_{ij}^{(1)}$, and the nonlinearity comes from the nonlinear terms in the polarization, P_i^{NL}, which depends on E^2 and higher orders of $\vec{E}$. It is worth mentioning that all the quantities depend on position and time, $\vec{E}(z,t)$, $\vec{p}(z,t)$, etc...

The general strategy for solving the wave equation is

1. We assume $\vec{E} = \sum_n \vec{E}_n$, which includes incident and outgoing fields. Each index n represent a field of one fourier component.

2. We make the fields approximately plane waves,

$$\vec{E}_n = \left[\frac{1}{2}A_n(z,\rho)e^{i(k_n z - \omega_n t)} + \frac{1}{2}A_n^*(z,\rho)e^{-i(k_n z - \omega_n t)}\right]\hat{\rho}. \tag{3.5}$$

The plane wave propagates along z and is cylindrically symmetric, and ρ is the only transverse coordinate. The direction of the polarization of the field is also $\hat{\rho}$, which implies that it is a transverse field, $\hat{\rho} \perp \hat{z}$. In addition, we assume that the slowly varying amplitude approximation holds, which means the amplitude may vary but not quickly with z and ρ. The amplitude $A_n(z,\rho)$ depends on the propagation distance z and the transverse direction ρ.

For example, in the case of second harmonic generation, the second harmonic field can grow with z, which is included in the z-dependence of A_n. Another example is self focusing, where in this case the beam width will change as it propagates. A_n gives us the shape of the beam as a function of both ρ and z.

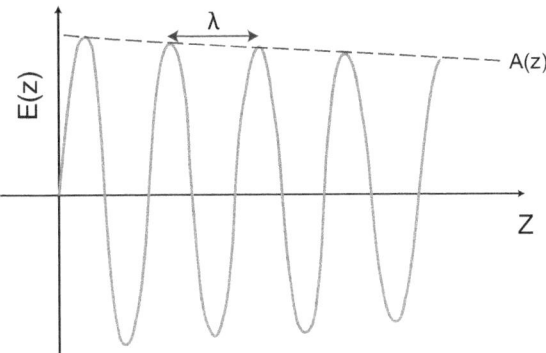

Figure 3.1: Slowly varying envelope approximation: the variation of $A(z)$ with respect to z is much smaller than the wavelength of the field's oscillation.

The slowly varying condition is applied to Maxwell's equations when differentiating the field with respect to the position. The derivative of A_n with respect to z should be much smaller than the derivative of $\exp[i(k_n z - \omega_n t)]$, or

$$\frac{\partial A_n}{\partial z} \ll A_n k_n. \tag{3.6}$$

As it is shown in Figure 3.1 the electric field oscillates with position and the envelope of the oscillation is the amplitude A_n, which changes very slowly on the scale of the wavelength of the oscillation.

Similarly,

$$\frac{\partial A_n}{\partial \rho} \ll A_n k_n, \tag{3.7}$$

where in this case the beam's amplitude changes negligibly in the transverse direction on the scale of the light's wavelength. Basically, when the amplitude is approximately constant over oscillations, the plane wave approximation holds. In terms of ρ, when the width of the wave is much larger than the wavelength, the plane wave approximation can be used.

3. Project out the fourier component of Maxwell's wave equation and sequentially keep terms of the same order in the electric field strength. We select only one frequency by fourier transformation and make the terms of the same order equal to each other.

$$\int dt e^{i\omega_n t} \left[\nabla \times \nabla \times \vec{E} + \frac{1}{c^2} \frac{\partial^2 \vec{D}}{\partial t^2} \right] = 0. \tag{3.8}$$

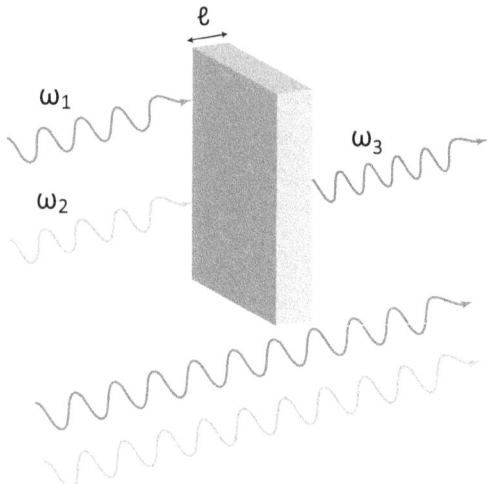

Figure 3.2: Sum frequency generation: the nonlinear interaction of two incident waves of frequencies ω_1 and ω_2 with a sample of with thickness l results in the generation of a wave of frequency $\omega_3 = \omega_1 + \omega_2$.

4. Simplify the expression

$$\nabla \times \nabla \times \vec{E} = -\nabla^2 \vec{E} + \nabla(\nabla \cdot \vec{E}). \qquad (3.9)$$

If A_n is constant along the direction of polarization, $\hat{\rho}$, then $\nabla \cdot \vec{E} = 0$, and we can ignore the second term of the right hand side of Equation 3.9.

3.2 Sum Frequency Generation - Non-Depletion Regime

Now we apply the same technique that we established for sum frequency generation in the non-depletion regime. In this regime, the sum frequency generation process can be described as two driving waves with frequencies ω_1 and ω_2, and a generated wave of frequency $\omega_3 = \omega_1 + \omega_2$. However, the intensity of the generated beam is so small that the initial beam is not depleted. Therefore, we can assume that the incident beams are unchanged, so they propagate as they would in a linear material. A schematic representation of the process is shown in Figure 3.2.

The electric field amplitudes are assumed to be independent of z, and are of the form,

$$\vec{E}_n = \left[\frac{1}{2}A_n(z)e^{i(k_n z - \omega_n t)} + \frac{1}{2}A_n^*(z)e^{-i(k_n z - \omega_n t)}\right]\hat{\rho}, \quad (3.10)$$

where the amplitudes are independent of ρ, and therefore represent infinitely-wide gaussian beams. To further simplify this case, we assume that all quantities, aside from wave-propagation along z, depend only on ρ; therefore, they are scalars, and the wave equation becomes,

$$-\nabla^2 E + \frac{\epsilon}{c^2}\frac{\partial^2 E}{\partial t^2} = -\frac{4\pi}{c^2}\frac{\partial^2 P^{NL}}{\partial t^2}. \quad (3.11)$$

In the non-depletion regime, the only nonlinear polarization that leads to sum frequency generation is,

$$\left(P^{NL}\right)^{\omega_1 + \omega_2} = \chi^{(2)} E_1^{\omega_1} E_2^{\omega_2}. \quad (3.12)$$

The ω_1 and ω_2 components of P^{NL} vanish. The wave equation of the fundamental beams, therefore, is linear, and the only nonlinear component is at frequency $\omega_3 = \omega_1 + \omega_2$. The solution to the linear wave equation (Equation 3.4) is a plane wave with $k_1 = \frac{\omega_1}{c}n_1$ and $k_2 = \frac{\omega_2}{c}n_2$. The zeroth order solution for ω_3 is also a plane wave of the same form, since $\chi^{(2)}$ is small for small P^{NL},

$$E_j = \frac{A_j}{2}e^{-i(k_j z - \omega_j t)}, \quad j = 1, 2, 3. \quad (3.13)$$

To get the first order correction at frequency ω_3, we substitute the sum of the incoming and outgoing fields into the wave equation (Equation 3.11) and project by Fourier trnasform the ω_3 components, which yields,

$$\frac{ik_3}{2}\frac{\partial A_3}{\partial z}e^{-i(k_3 z - \omega_3 t)} + \frac{k_3^2}{2}A_3(z)e^{-i(k_3 z - \omega_3 t)} - \frac{\epsilon_3 \omega_3^2}{2c^2}A_3 e^{-i(k_3 z - \omega_3 t)}$$
$$= \frac{\pi \omega_3^2}{c^2}\chi^{(2)}(-\omega_3; \omega_1, \omega_2) A_1 e^{-i(k_1 z - \omega_1 t)} A_2 e^{-i(k_2 z - \omega_2 t)}.$$
$$(3.14)$$

Since we have used the slowly varying envelope approximation, the terms with second derivative or higher with respect to z are negligible.

To solve for the zeroth order equations, when $\chi^{(2)} = 0$, the right hand side of Equation 3.14 is zero. Additionally the first term on the left hand side is also zero because, to zeroth order, there will be no conversion of the intensity of the $\omega_1 + \omega_2$ wave to other colors due to nonlinearty. The zeroth-order terms in Equation 3.14 yield,

$$k_3^2 = \frac{\epsilon_3 \omega_3^2}{c^2} = \left(1 + 4\pi \chi^{(1)}\right) \frac{\omega_3^2}{c^2}. \tag{3.15}$$

Using the fact that $n^2 = \epsilon$, Equation 3.15 becomes,

$$k_3 = \frac{n_3 \omega_3}{c}. \tag{3.16}$$

Equation 3.16 indicates that all the zeroth-order solutions of the wave equation propagate with $k_i = \frac{n_i \omega_i}{c}$, where $i = 1, 2, 3$. As previously discussed, in the non-depletion regime there are no nonlinear effects acting on the fields at frequencies ω_1 and ω_2. Thus, the exact solution for those two frequencies is simply a linear wave as given by Equation 3.13. The solution of the wave equation for the ω_3 wave is the superposition of the linearly-propagating wave and the next order correction. The first-order correction can be calculated from the two first-order correction terms, which we previously ignored in the Equation 3.14, which yields

$$\frac{ik_3}{2} \frac{\partial A_3}{\partial z} e^{-i(k_3 z - \omega_3 t)} = \frac{\pi \omega_3^2}{c^2} \chi^{(2)}(-\omega_3; \omega_1, \omega_2)$$
$$A_1 e^{-i(k_1 z - \omega_1 t)} A_2 e^{-i(k_2 z - \omega_2 t)}. \tag{3.17}$$

The fourier transform of Equation 3.17 is obtained by multiplying it by $e^{-i\omega_3 t}$ and integrating over time (one should be careful that $k_3 \neq k_1 + k_2$). This yields

$$\frac{ik_3}{2} \frac{\partial A_3}{\partial z} e^{-ik_3 z} = \frac{\pi k_3^2}{\epsilon_3} \chi^{(2)}(-\omega_3; \omega_1, \omega_2) A_1 e^{-ik_1 z} A_2 e^{-ik_2 z}, \tag{3.18}$$

therefore,

$$\frac{\partial A_3}{\partial z} = \frac{2\pi k_3}{i\epsilon_3} \chi^{(2)}(-\omega_3; \omega_1, \omega_2) A_1 A_2 e^{-i\Delta k z}, \tag{3.19}$$

where $\Delta k = k_1 + k_2 - k_3$ is called the phase mismatch.

In the non-depletion regime the coefficients of the exponentials, A_1 and A_2 are constant. Thus, it is straightforward to integrate Equation 3.19. Using the geometry shown in Figure 3.2, Equation 3.19 yields

$$A_3 = \frac{2\pi k_3 \chi^{(2)}(-\omega_3; \omega_1, \omega_2)}{\epsilon_3 \Delta k} A_1 A_2 \left[e^{-i\Delta k l} - 1\right]. \tag{3.20}$$

Lecture Notes in Nonlinear Optics

Equation 3.20 represents that field amplitude of the ω_3 wave at the output plane of the material located at $z = l$. Light at frequency ω_3 is generated by the nonlinear interaction of light at frequencies ω_1 and ω_2. But there are lots of photons with frequencies ω_1 and ω_2 that do not interact within the material and they pass right through the sample unchanged.

To simplify Equation 3.20 we write it as follows,

$$A_3 = -\frac{4i\pi k_3 \chi^{(2)} A_1 A_2}{\epsilon_3 \Delta k} \sin\left(\frac{\Delta k l}{2}\right) \exp\left(\frac{-i\Delta k l}{2}\right). \tag{3.21}$$

Since we are interested in measurable quantities, it is useful to determine the intensity from the field. The intensity is calculated using Poynting's theorem, which in gaussian units is of the form,

$$I_i = \frac{c}{8\pi} n_i |A_i|^2. \qquad i = 1, 2, 3 \tag{3.22}$$

Given this definition, we can determine I_3 by substituting Equation 3.21 into Equation 3.22,

$$I_3 = \frac{c}{8\pi} n_3 |A_3|^2 = \frac{cn_3}{8\pi} \left(\frac{4\pi k_3 \chi^{(2)} A_1 A_2}{\epsilon_3 \Delta k}\right)^2 \sin^2\left(\frac{\Delta k l}{2}\right). \tag{3.23}$$

Replacing all values of $|A_i|^2$ with $8\pi I_i / c n_i$ yields

$$I_3 = \frac{32\pi^3 \omega_3^2}{c^3} \left(\chi^{(2)}\right)^2 \frac{I_1 I_2}{n_1 n_2 n_3} L_c^2 \sin^2\left(\frac{l}{L_c}\right), \tag{3.24}$$

where $L_c = 2/\Delta k$ is called the coherence length. Equation 3.24 shows that the output intensity at the sum frequency oscillates as a function of the sample thickness over a coherence length. The oscillation in the intensity is produced by the phase mismatch between the three waves that are traveling at different speeds, $v_i = c/n_i$ and causing interference. In order to get the most efficient from sum frequency generation, the waves ω_1 and ω_2 should travel with a phase velocity such that the wave produced at frequency ω_3 at one point in the material is always in phase with the ω_3 wave, which has already been generated up stream.

There are two conditions to be considered in Equation 3.24, which are shown in Figure 3.3. First, the intensity of the generated wave depends on the intensity of the incident waves, and on $\chi^{(2)}$; so, the higher the intensity of the initial waves or the larger the $\chi^{(2)}$, the higher the intensity of the generated wave (solid red line). Secondly, phase matching in the non-depletion limit with $\Delta k \to 0$ leads to $\sin^2(l/L_c) \to (l/L_c)^2$; so, $I_3 \propto l^2$, which is shown

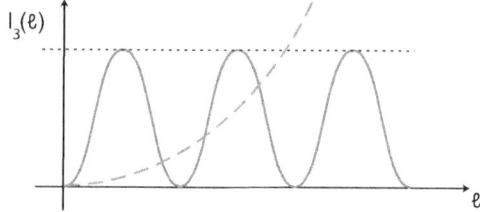

Figure 3.3: The intensity of generated light as a function of sample length, l, is proportional to $A_1 A_2 \chi^{(2)}$ and a small-amplitude oscillating function in the non-depletion regime (solid red curve) and in the phase-matching condition (the dashed green curve), it is proportional to l^2.

as the dashed green curve in Figure 3.3. For the non-depletion limit to hold, the amount of light being generated should be very small compared to the intensity of the incident waves. Thus, in the case of no phase matching, the intensity of the light that is incident on the sample should be small; and, in the second case the sample should be thin enough to not create an appreciable amount of second harmonic light.

3.3 Sum Frequency Generation - Small Depletion Regime

Now we consider sum frequency generation in the small depletion regime, relaxing the condition that the incident beams are not depleted. As shown in Figure 3.4, two incident photons of frequencies ω_1 and ω_2, and amplitudes A_1 and A_2 interact to produce a photon at frequency ω_3. Downstream, the photon at frequency ω_3 interacts with another photon of frequency ω_2 to produce a photon of frequency ω_1. Similarly, a photon at ω_2 could be produced. In the no depletion approximation, not enough light is generated at frequency ω_3 to to lead to back conversion; but, in the small depletion regime, ω_3 can convert back to ω_1 by mixing with ω_2.

Before proceeding, it is interesting to point out that the back-conversion process appears to violate energy conservation. This issue can be be resolved by recognizing the actual process shown in the inset of the Figure 3.4. Here, the photon at frequency ω_2 stimulates the emission of another photon with frequency ω_2 through the annihilation of a photon of frequency ω_3. The original photon of frequency ω_2 passes through the sample without any changes.

Lecture Notes in Nonlinear Optics

In this process energy is conserved. So, energy conservation appears to be violated because not all the photons are shown in the diagram. However, using such diagrams that appear to violate energy conservation yield the correct mixing efficiency.

We begin with by rewriting Equation 3.19 and its complex conjugate,

$$\frac{\partial A_3}{\partial z} = -\frac{2\pi i k_3}{\epsilon_3}\chi^{(2)}(-\omega_3;\omega_1,\omega_2)A_1 A_2 e^{-i(k_1+k_2-k_3)z}, \quad (3.25)$$

and,

$$\frac{\partial A_3^*}{\partial z} = \frac{2\pi i k_3}{\epsilon_3}\chi^{(2)}(\omega_3;-\omega_1,-\omega_2)A_1^* A_2^* e^{-i(-k_1-k_2+k_3)z}. \quad (3.26)$$

Here we used the fact that complex conjugate of the response function yields a change of sign in the arguments. With this notation it is easier to understand how ω_1 and ω_2 combine to give ω_3. The outgoing wave is always accompanied with a minus sign in the argument, and the positive frequency means the wave is incident on the material. This is consistent with the processes shown in Figure 3.4.

We can equivalently write $\chi^{(2)}(\omega_3;-\omega_1,-\omega_2)$ as $\chi^{(2)}(-\omega_1;\omega_3,-\omega_2)$ to rewrite Equation 3.26 as

$$\frac{\partial A_3^*}{\partial z} = \frac{2\pi i k_3}{\epsilon_3}\chi^{(2)}(-\omega_1;\omega_3,-\omega_2)A_1^* A_2^* e^{-i(-k_1-k_2+k_3)z}. \quad (3.27)$$

The signs of the frequencies in the argument of $\chi^{(2)}$ clearly match the signs of k_i in the exponential. This conjugate equation represents the process where ω_3 is being depleted. We can write this equation for A_1 and A_2 as well,

$$\frac{\partial A_1}{\partial z} = -\frac{2\pi i k_1}{\epsilon_1}\chi^{(2)}(-\omega_1;\omega_3,-\omega_2)A_2^* A_3 e^{-i(-k_1-k_2+k_3)z}, \quad (3.28)$$

and

$$\frac{\partial A_2}{\partial z} = -\frac{2\pi i k_2}{\epsilon_2}\chi^{(2)}(-\omega_2;\omega_3,-\omega_1)A_1^* A_3 e^{-i(-k_1-k_2+k_3)z}. \quad (3.29)$$

As one may notice, the susceptibilities are the same for Equations 3.27, 3.28, and 3.29. This allows us to derive a relationship between the derivatives of the intensities by using the derivative of amplitudes in the small depletion regime. Differentiating Equation 3.22 with respect to z yields,

$$\frac{\partial I_i}{\partial z} = \frac{c n_i}{8\pi}\frac{\partial |A_i|^2}{\partial z} = \frac{c n_i}{8\pi}\left(\frac{\partial A_i}{\partial z}A_i^* + \frac{\partial A_i^*}{\partial z}A_i\right). \quad i=1,2,3. \quad (3.30)$$

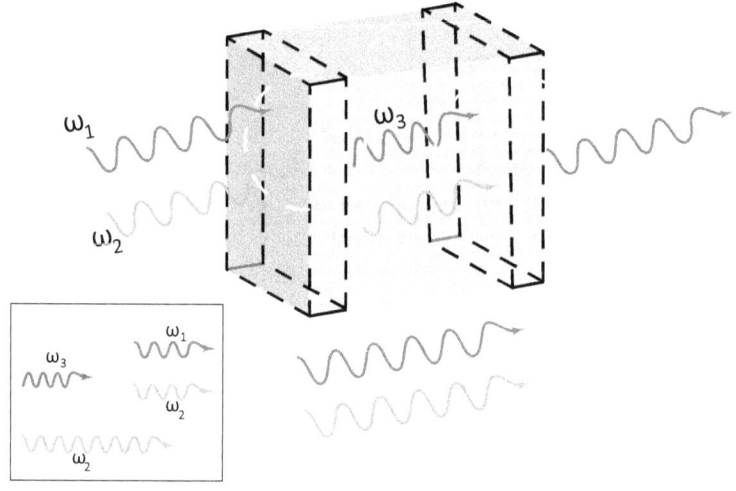

Figure 3.4: Sum frequency generation in the small depletion regime is represented by the nonlinear interaction of two incident waves with frequencies ω_1 and ω_2 within the material, to generate a photon of frequency ω_3. Downstream, this generated photon and a photon of frequency ω_2 interact and create a photon of frequency ω_1. The inset shows how the energy is conserved in the process of the destruction of a photon of frequency ω_3 and creation of two photons of frequencies ω_1 and ω_2.

Lecture Notes in Nonlinear Optics 63

Next, we substitute A_i and $\partial A_i/\partial z$ and their complex conjugates into Equation 3.30 for each of the intensities, using Equations 3.27, 3.28, and 3.29. Before proceeding we must impose additional assumptions. First, we assume that all light frequencies are far from the resonance frequencies of the material. Therefore, $\chi^{(2)}$ is real. Secondly, we assume full permutation symmetry, which implies that $\chi^{(2)}$ is independent of the wavelength. Then $\chi^{(2)}$ is independent of frequency, allowing it to be removed as a common factor. Finally, we assume the refractive indices are real. Taking all these considerations into account results in,

$$\frac{\partial I_1}{\partial z} = \frac{cn_1}{8\pi}\left(-\frac{2\pi i k_1 \chi^{(2)}}{\epsilon_1} A_2^* A_3 A_1^* e^{i\Delta k z} + \frac{2\pi i k_1 \chi^{(2)}}{\epsilon_1} A_2 A_3^* A_1 e^{-i\Delta k z}\right), \quad (3.31)$$

$$\frac{\partial I_2}{\partial z} = \frac{cn_2}{4}\frac{ik_2 \chi^{(2)}}{\epsilon_2}\left(-A_1^* A_3 A_2^* e^{i\Delta k z} + A_1 A_3^* A_2 e^{-i\Delta k z}\right), \quad (3.32)$$

and

$$\frac{\partial I_3}{\partial z} = \frac{cn_3}{4}\frac{ik_3 \chi^{(2)}}{\epsilon_3}\left(-A_1 A_2 A_3^* e^{-i\Delta k z} + A_1^* A_2^* A_3 e^{i\Delta k z}\right). \quad (3.33)$$

In Equations 3.31, 3.32, and 3.33, the first term represents the depletion of the photon with specified frequency into other frequencies; and, the second term shows generation of that frequency. To simplify the intensity derivatives, we can take advantage of the dispersion relations, $k = n\omega/c$ and $\epsilon = n^2$, and rewrite Equations 3.31 - 3.33 as follow,

$$\frac{\partial I_1}{\partial z} = \frac{i\omega_1 \chi^{(2)}}{4}\left(-A_2^* A_3 A_1^* e^{i\Delta k z} + A_2 A_3^* A_1 e^{-i\Delta k z}\right), \quad (3.34)$$

$$\frac{\partial I_2}{\partial z} = \frac{i\omega_2 \chi^{(2)}}{4}\left(-A_1^* A_3 A_2^* e^{i\Delta k z} + A_1 A_3^* A_2 e^{-i\Delta k z}\right), \quad (3.35)$$

and

$$\frac{\partial I_3}{\partial z} = \frac{i\omega_3 \chi^{(2)}}{4}\left(-A_1 A_2 A_3^* e^{-i\Delta k z} + A_1^* A_2^* A_3 e^{i\Delta k z}\right). \quad (3.36)$$

Adding these equations together, gives

$$\begin{aligned}\frac{\partial I_1}{\partial z}+\frac{\partial I_2}{\partial z}+\frac{\partial I_3}{\partial z} &= \frac{i\chi^{(2)}}{4}\Big[e^{-i\Delta kz} \\ &\quad\times (\omega_1 A_2 A_3^* A_1 + \omega_2 A_1 A_3^* A_2 - \omega_3 A_1 A_2 A_3^*) \\ &\quad + e^{i\Delta kz}(-\omega_1 A_2^* A_3 A_1^* - \omega_2 A_1^* A_3 A_2^* + \omega_3 A_1^* A_2^* A_3)\Big] \\ &= \frac{i\chi^{(2)}}{4}\left(e^{i\Delta kz}-e^{-i\Delta kz}\right)(\omega_3-\omega_1-\omega_2) \\ &\quad \times [A_1^* A_2^* A_3 + A_1 A_2 A_3^*].\end{aligned}$$

(3.37)

Using the fact that $\omega_3 = \omega_1 + \omega_2$,

$$\frac{\partial I_1}{\partial z}+\frac{\partial I_2}{\partial z}+\frac{\partial I_3}{\partial z} = 0, \tag{3.38}$$

which implies

$$I_1 + I_2 + I_3 = \text{constant}. \tag{3.39}$$

This is a simple restatement of the conservation of the energy. In the case of sum frequency generation, the total intensities remains unchanged with a value of the initial intensity of the two input beams, $I_1(z=0)+I_2(z=0)$.

Superposition holds for the linear Maxwell's equations, so we can express the total electric field as a sum over all fields that are present. In a nonlinear equation, superposition doesn't hold; but, it approximately holds and so the energy should be approximately conserved. However, it may not come out exactly that way because of all the approximations used in solving the problem. Thus we should directly check for conservation of energy. The energy might not be a constant and it may flow to higher order terms that we are not taking into account. The other issue is that we only considered the energy of the fields. If we are on resonance we also have to include the energy of the material. The energy of the fields might not be constant separately and the energy can flow in or out of the molecules. The total energy of the fields and the molecules therefore should be constant. Thus we should be careful when applying these results to situations where the approximations we made may not hold.

Lecture Notes in Nonlinear Optics

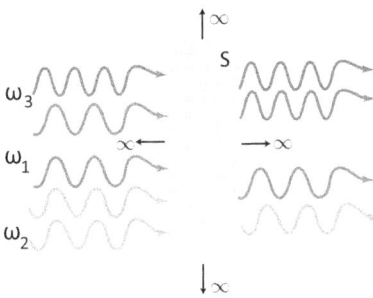

Figure 3.5: The Manley-Rowe equation expresses the fact that the absolute change in the number of photons at every frequency is the same in each interaction. For example, the destruction of one photon at ω_1 and one photon at ω_2 yields one photon created at frequency ω_3.

3.4 Aside - Physical Interpretation of the Manley-Rowe Equation

Let's revisit Equations 3.34 through 3.36. If we divide the derivative of each intensity by its frequency, the result will be the same for all frequencies, but only differs in the sign. It will be positive for the incoming waves and negative for the outgoing waves, or,

$$\frac{1}{\omega_1}\frac{\partial I_1}{\partial z} = \frac{1}{\omega_2}\frac{\partial I_2}{\partial z} = -\frac{1}{\omega_3}\frac{\partial I_3}{\partial z}. \qquad (3.40)$$

Equation 3.40 is called the Manley-Rowe expressions and they are related to energy conservation.

To understand this relationship more deeply, let's consider the process shown in Figure 3.5. We integrate the Manley-Rowe equations over a surface that is perpendicular to the beam propagation direction, and goes to infinity in all directions is a region where all three beams, I_1, I_2, and I_3 are present,

$$\frac{\partial}{\partial z}\left[\frac{1}{\omega_1}\int I_1 ds = \frac{1}{\omega_2}\int I_2 ds = -\frac{1}{\omega_3}\int I_3 ds\right]. \qquad (3.41)$$

It is useful to remember the units of the intensity, $\left[\frac{\text{Power}}{\text{Area}}\right]$. The integration of the intensity over area gives power, $\left[\frac{\text{Energy}}{\text{Time}}\right]$. On the other hand if we divide $1/\omega$ by $\hbar$, $\hbar\omega$ in the denominator also has dimensions of energy. Thus, the energy passing through the surface divided by the energy per photon $\hbar\omega$

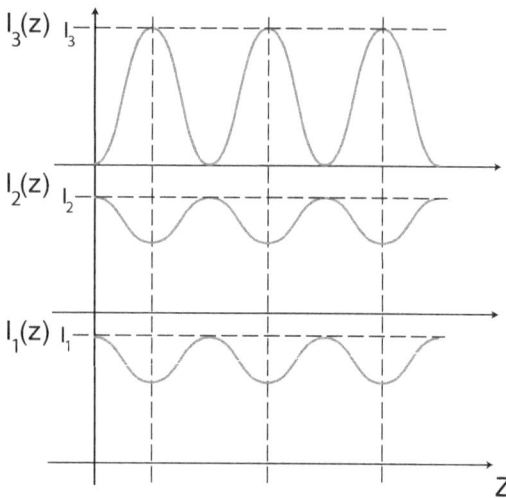

Figure 3.6: Comparison of the intensities of the three beams of frequency ω_1, ω_2, and ω_3. in the sum frequency generation process. I_1 and I_2 have their minima when I_3 peaks.

is simply the number of photons passing through the surface. Equation 3.41 thus indicates that the change in the number of photons in the first beam is equal to the change in the number of photons in the second beam and the change in the number of photons in the third beam with a minus sign, or

$$\Delta N_1 = \Delta N_2 = -\Delta N_3. \tag{3.42}$$

This is exactly the picture we have been using: two photons are destroyed at frequencies ω_1 and ω_2, and a photon is created at frequency ω_3. These results in the small depletion regime are illustrated in Figure 3.6. I_1 and I_2 have high intensity and can be depleted, and wherever these two are minimum, I_3 is peaked.

3.5 Sum Frequency Generation with Depletion of One Input Beam

Previously, we considered the process of sum frequency generation for two undepleted beams. We now turn to the case where one of the two beams is allowed to be depleted. In this particular case the undepleted beam, ω_1, will

Lecture Notes in Nonlinear Optics 67

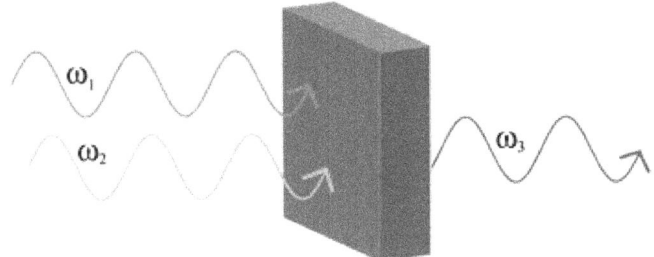

Figure 3.7: The idler, ω_1, and the pump, ω_2, undergo sum frequency generation in a nonlinear crystal and create an output with frequency $\omega_3 = \omega_1 + \omega_2$.

be referred to as the pump and the depleted beam, ω_2, will be referred to as the idler. The process is shown in figure 3.7.

There are four characteristics of the assumptions underlying the calculation of sum frequency generation with depleted idler:

1. Strong undepleted pump at frequency ω_2.

2. Weak idler at frequency ω_1 which can be fully depleted.

3. The sum frequency and probe intensity are much smaller than the pump, $I_1 + I_3 \ll I_2$ where subscripts denote the frequencies ω_1, ω_2, and ω_3.

4. The phase matching condition holds, therefore, $\Delta k = 0$.

These conditions are easily met experimentally. The previous sections have already defined the amplitudes and coordinates that we use here. The intensity of the ith beam, I_i, is related to its amplitude, A, according to,

$$I_i = \frac{cn_i}{8\pi}|A_i|^2. \tag{3.43}$$

Also, as previously shown in the derivation of Equation 3.19,

$$\frac{dA_j}{dz} = \frac{4\pi}{in_j}\frac{\omega_j}{c}\chi^{(2)}A_3 A_l^* e^{i\Delta kz}. \tag{3.44}$$

The processes we are considering are the cases defined by $j = 1$ and $l = 2$; and, $j = 2$ and $l = 1$. Also, note that the phase matching condition implies $\exp[i\Delta kz] = 1$.

Because there is no depletion of the pump, Equation 3.44 implies that $\partial A_2/\partial z = 0$. Therefore, A_2 is a constant. So, considering only A_1 and A_3, we get

$$k_1 = \frac{4\pi}{in_1} \frac{\omega_1}{c} \chi^{(2)} A_2^*, \tag{3.45}$$

where k_1 is a constant. Likewise,

$$k_3 = \frac{4\pi}{in_3} \frac{\omega_3}{c} \chi^{(2)} A_2, \tag{3.46}$$

where k_3 is also a constant. Therefore, Equation 2 can be expressed as,

$$\boxed{\frac{dA_1}{dz} = k_1 A_3.} \tag{3.47}$$

$$\boxed{\frac{dA_3}{dz} = k_3 A_1.} \tag{3.48}$$

Equations 3.47 and 3.48 can now be combined to write the second spatial derivative of A_3 with respect to the $\hat{z}$ direction. This gives,

$$\frac{d}{dz}\left(\frac{1}{k_3}\frac{dA_3}{dz}\right) = k_1 A_3, \tag{3.49}$$

and therefore,

$$\frac{d^2 A_3}{dz^2} = k_1 k_3 A_3. \tag{3.50}$$

If we assume that we are off resonance, then $\chi^{(2)}$ is real. Therefore, $k_1 k_2 = -K^2$, and so the resulting one-dimensional Helmholtz equation can be easily solved. So, with the resulting equation,

$$\frac{d^2 A_3}{dz^2} = -K^2 A_3, \tag{3.51}$$

the solution is in the form of sines and cosines with the two constants to be evaluated using the proper boundary conditions of the experiment. The solution of A_3 can then be written as

$$A_3(z) = B\cos(Kz) + C\sin(Kz). \tag{3.52}$$

The solution to A_1 is found by substituting Equation 3.52 into Equation 3.48, which yields,

$$A_1(z) = \frac{-BK}{k_3}\sin(Kz) + \frac{CK}{k_3}\cos(Kz). \tag{3.53}$$

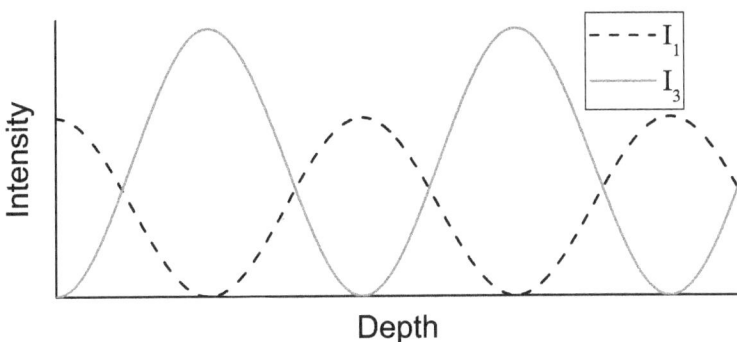

Figure 3.8: The intensities at frequencies ω_1 and ω_3 plotted as a function of propagation distance.

Since, A_2 is a constant, we express it in complex form,

$$A_2 = A_{20} e^{i\phi_2}, \tag{3.54}$$

where A_{20} is the amplitude and ϕ_2 the phase.

When $z = 0$, $A_1(z = 0)$ is at the input interface of the nonlinear crystal. The intensity of the beam at frequency ω_1 is maximum at the input interface. Therefore, this boundary condition yields,

$$A_1(z) = A_1(0)\cos(Kz) \tag{3.55}$$

To evaluate the integration constant for A_3, we recognize that $\frac{k_3}{K} = \frac{\sqrt{k_3}}{\sqrt{k_1}}$, which using Equations 3.45 and 3.46 gives,

$$\frac{k_3}{K} = \sqrt{\frac{n_1 \omega_3}{n_3 \omega_1} \frac{A_2}{A_2^*}}. \tag{3.56}$$

Using Equation 3.54, we recognize that $\sqrt{A_2/A_2^*}$ is the phase, so Equation 3.56 becomes,

$$\frac{k_3}{K} = -i e^{i\phi_2} \sqrt{\frac{n_1 \omega_3}{n_3 \omega_1}}. \tag{3.57}$$

Using the boundary condition that $A_3(0) = 0$ will lead to $B = 0$ in Equation 3.52. Substituting Equation 3.52 into Equation 3.51 with the help of Equation 3.57 yields,

$$A_3(z) = -ie^{i\phi_2}\sqrt{\frac{n_1}{n_3}\frac{\omega_3}{\omega_1}}A_1(0)\sin(Kz). \quad (3.58)$$

The intensity of each field is calculated using Equation 3.43, and is proportional to the square of the amplitude of the waves as they propagate through the nonlinear crystal. Thus the intensities are given by,

$$\boxed{I_1 = \frac{cn_1}{8\pi}|A_1(0)|^2\cos^2(Kz).} \quad (3.59)$$

$$\boxed{I_3 = \frac{cn_3}{8\pi}\left[|A_1(0)|^2\frac{n_1}{n_3}\frac{\omega_3}{\omega_1}\sin^2(Kz)\right].} \quad (3.60)$$

The intensities as a function of z are shown in Figure 3.8.

3.6 Difference Frequency Generation

Next we consider difference frequency generation. The difference beam at frequency $\omega_3 = |\omega_1 - \omega_2|$ is generated by a strong pump beam at frequency ω_1 and the amplified beam (described later) at frequency ω_2. The process is shown by Figure 3.9.

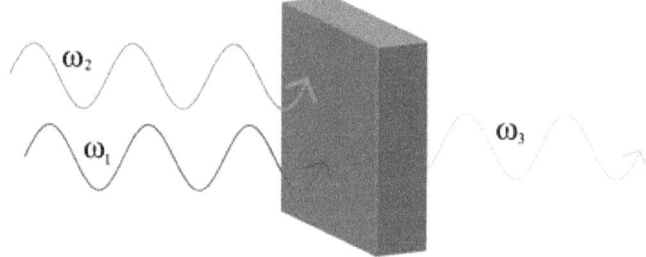

Figure 3.9: The pump beam at frequency ω_1 and the amplified beam at frequency ω_2 undergo difference frequency generation in a nonlinear crystal. This creates the difference beam at frequency $\omega_3 = \omega_1 - \omega_2$.

The simplifying assumptions used in the following derivation are as follows:

1. Strong undepleted pump at frequency ω_1.

Lecture Notes in Nonlinear Optics 71

2. The beam frequencies are far from material resonances so there is full permutation symmetry.

3. $\chi^{(2)}$ is independent of frequency.

4. The phase matching condition holds, therefore, $\Delta k = 0$.

As in the sum frequency generation calculations in Sections 3.1 and 3.2, Maxwell's equations are reduced to,

$$\frac{dA_2}{dz} = 4\pi \frac{\omega_2^2}{c^2} \frac{1}{ik_2} \chi^{(2)} A_1 A_3^*, \tag{3.61}$$

since A is a constant in the undepleted pump approximation. Defining the constant K_2 as,

$$K_2 = 4\pi \frac{\omega_2^2}{c^2} \frac{1}{ik_2} \chi^{(2)} A_1, \tag{3.62}$$

Equation 3.61 can then be rewritten in the form

$$\frac{dA_2}{dz} = K_2 A_3^*. \tag{3.63}$$

In parallel to Section 3.5, the spatial derivative of A_3 that results from Maxwell's equations yields,

$$\frac{dA_3}{dz} = 4\pi \frac{\omega_3^2}{c^2} \frac{1}{ik_3} \chi^{(2)} A_1 A_2^*. \tag{3.64}$$

Since A is a constant, we define the constant K_3 to be

$$K_3 = 4\pi \frac{\omega_3^2}{c^2} \frac{1}{ik_3} \chi^{(2)} A_1. \tag{3.65}$$

Therefore, Equation 3.64 can be rewritten in the form

$$\frac{dA_3}{dz} = K_3 A_2^*. \tag{3.66}$$

Equations 3.63 and 3.66 yield

$$\frac{d^2 A_2}{dz^2} = K_2 \frac{dA_3^*}{dz}, \tag{3.67}$$

which can be rewritten as

$$\frac{d^2 A_2}{dz^2} = K_2 K_3^* A_2. \tag{3.68}$$

Equation 3.68 is in the form of Equation 3.50 except that $K_2 K_3^* > 0$. Setting $K^2 = K_2 K_3^*$, the solution becomes,

$$A_2(z) = C \sinh(Kz) + D \cosh(Kz). \tag{3.69}$$

Using Equation 3.66 and Equation 3.69, A_3 is found to be of the form,

$$A_3(z) = \frac{-K}{K_2^*}[C \cosh(Kz) + D \sinh(Kz)]. \tag{3.70}$$

$z = 0$, corresponds to the input interface of the sample, so $A_3(z=0) = 0$ and $A_2(z=0) \neq 0$. These boundary conditions imply that $C = 0$ and $D = A_2(0)$, so

$$A_2(z) = A_2(0) \cosh(Kz). \tag{3.71}$$

By inserting Equation 3.71 into Equation 3.63 we get,

$$A_3(z) = -i \frac{A_1}{|A_1|} A_2(0) \sqrt{\frac{k_3 \, n_2}{k_2 \, n_3}} \sinh(Kz). \tag{3.72}$$

It should be noted that A_1 is assumed undepleted and is considered to be constant, which was used in the solution of A_2 and A_3.

The intensities can then be written as

$$\boxed{I_2 = \frac{c n_2}{8\pi} \left[|A_2(0)|^2 \cosh^2(Kz) \right],} \tag{3.73}$$

$$\boxed{I_3 = \frac{c n_3}{8\pi} \left[|A_2(0)|^2 \frac{k_3 \, n_2}{k_2 \, n_3} \sinh^2(Kz) \right],} \tag{3.74}$$

and are plotted in Figure 3.10. Note that hyperbolic functions diverge as $z \to \infty$, so Equations 3.73 and 3.73 must only be used in the low-intensity regime. Note that the wave at frequency ω_2 is amplified in the process that generates the difference-frequency wave at frequency ω_3.

3.7 Second Harmonic Generation

We now consider second harmonic generation, a special case of sum frequency generation. Second harmonic generation can be described as a process that two incident light of the same frequency $\omega \equiv \omega_1$ interact with the material, generating an output light of frequency $2\omega \equiv \omega_2$, as shown in Figure 3.11. Under certain conditions, the intensity of the fundamental wave (which is the incident light) decreases and that of the second harmonic wave increases

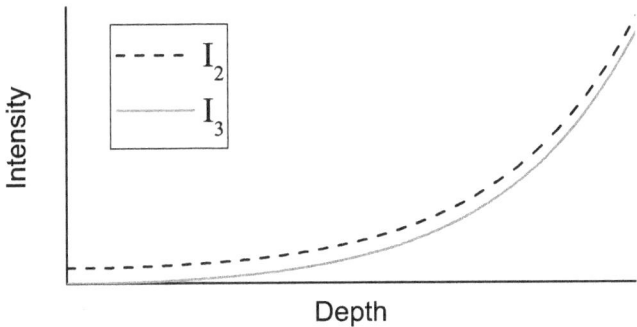

Figure 3.10: The intensities at frequencies ω_2 and ω_3 as a function of depth into the material.

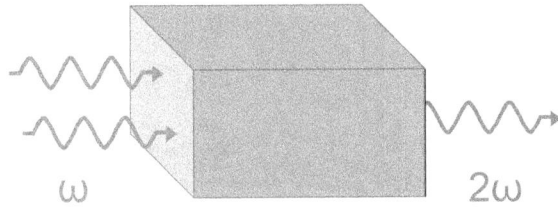

Figure 3.11: In second harmonic generation, two incident wave of frequency ω interact with a nonlinear medium, generating an output wave of frequency 2ω.

as two waves propagate in a lossless dielectric material (which is the case we are considering throughout our discursion), but the total intensity remains invariant.

As we did previously in sum frequency generation, starting from the Nonlinear Wave Equation 3.11, with the nonlinear polarization associated with second harmonic generation (similar to Equation 3.12),

$$\left(P^{NL}\right)^{2\omega} = \frac{1}{2}\chi^{(2)} E^{\omega 2}, \tag{3.75}$$

and

$$\left(P^{NL}\right)^{\omega+2\omega} = \chi^{(2)} E^{\omega} E^{2\omega}, \tag{3.76}$$

where E^{ω} and $E^{2\omega}$ are complex amplitudes of electric fields of the fundamental wave and the second harmonic wave, respectively. We obtain the coupled

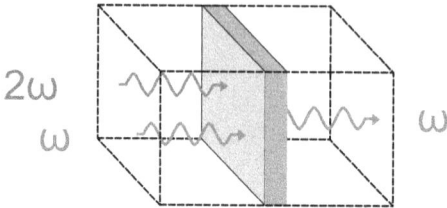

Figure 3.12: This process is difference frequency generation as we have seen in the previous section. The fundamental wave of frequency ω and the generated second harmonic wave of frequency 2ω interacts with the material and propagates through the material resulting in an output wave of frequency ω.

equations

$$\frac{dA_2}{dz} = \frac{2\pi}{in_2}\frac{\omega_2}{c}\chi^{(2)}A_1^2 e^{i\Delta k z}, \quad (3.77)$$

and

$$\frac{dA_1}{dz} = \frac{4\pi}{in_1}\frac{\omega_1}{c}\chi^{(2)}A_2 A_1^* e^{-i\Delta k z}, \quad (3.78)$$

where $\Delta k = 2k_1 - k_2$. Equation 3.77 represents the process of two fundamental waves interacting with a nonlinear medium and generating the second harmonic wave, as graphically shown in Figure 3.11. Unlike the non-depletion case, A_1 is no longer a constant but depends on z, the propagating distance. Once the second harmonic wave is generated, it may undergo the process of difference frequency generation by interacting with the material together with the fundamental wave, as mathematically presented in Equation 3.78 and graphically illustrated in Figure 3.12. If we rewrite these coupled equations in terms of intensities, the resulting solutions can be expressed by Jacobi elliptic functions. We will not solve these coupled equations here, but we will discuss the results under various conditions.

We first consider phase-matching condition, $\Delta k = 0$. The behavior of waves propagating through a material depends on initial conditions of the coupled equations. If two waves of frequency ω and 2ω are incident on a nonlinear medium, hence given the initial conditions that the intensity of the fundamental wave $I_1(z=0)$ and that of the second harmonic wave $I_2(z=0)$ both have nonzero values on the incident surface ($z=0$), I_1 and I_2 will oscillate while two waves propagate through the medium, but the total intensity is conserved according to Manley-Rowe relations. In other words, the fundamental wave and the second harmonic wave exchange energy when traveling in the medium as illustrated in Figure 3.13. The next case, which is the most

Lecture Notes in Nonlinear Optics 75

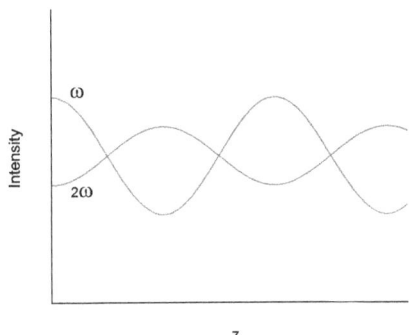

Figure 3.13: Intensities oscilate while two waves of of frequencies ω and 2ω propagate in the medium under phase-matching condition. This behavior is due to the initial non-zero incident intensities of both waves.

often used one, is to generate the second harmonic wave with only the fundamental wave incident on the medium; hence the initial conditions are now $I_1(z=0) \neq 0$ and $I_2(z=0) = 0$. The initial conditions imply $A_1(z=0) \neq 0$ and $A_2(z=0) = 0$ in the coupled equations, therefore we have $\left(\frac{dA_2}{dz}\right)_{z=0} \neq 0$ and $\left(\frac{dA_1}{dz}\right)_{z=0} = 0$. When the fundamental wave is incident on the material, the second harmonic wave is generated and becoming stronger as it propagates in the material. The energy gained by the second harmonic wave comes from the fundamental wave incident on the material, as shown in Figure 3.14.

In the case of phase-mismatching which occurs when $\Delta k \neq 0$ as solved by Armstrong et al., I_1 and I_2 oscillate as the waves propagate through the material. We are interested in second harmonic generation, therefore we examine the amplitude of the second harmonic wave as it travels through the material. With varying values of Δk as illustrated in Figure 3.15, if $\Delta k = 0$, the amplitude of the second harmonic wave gradually increases; if $\Delta k \neq 0$, the second harmonic wave can not be generated efficiently. Consequently phase-mismatching reduces the efficiency of converting energy to the second harmonic wave from the fundamental wave.

The solutions of coupled equations at the undepleted limit will give the same results as we saw previously in sum frequency generation under the same limit, with the incident waves of frequency ω and output wave of frequency 2ω. Therefore we have the intensity of the second harmonic wave

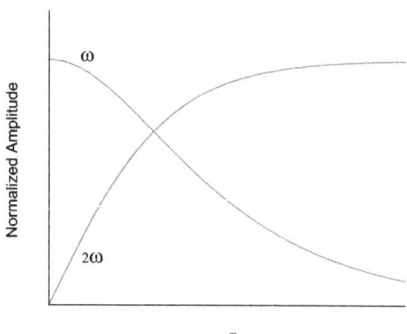

Figure 3.14: The energy is converted to the second harmonic wave from the fundamental wave. Notice that initially the slope of the fundamental wave is zero but that of the second harmonic wave begins with a non-zero value.

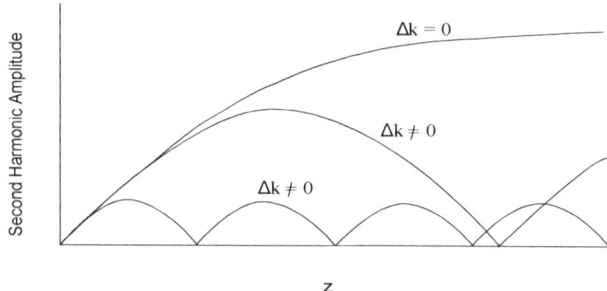

Figure 3.15: Phase-mismatching reduces the efficiency of generating second harmonic wave significantly.

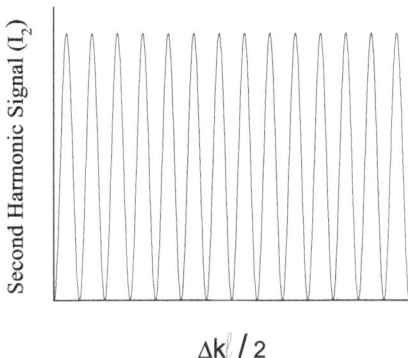

Figure 3.16: The undepleted limit solution of coupled equations.

acording to Equation 3.24

$$I_2 = \frac{32\pi^3 \omega_2^2}{c^3} \left(\chi^{(2)}\right)^2 \frac{I_1^2}{n_1^2 n_2} l^2 \text{sinc}^2 \left(\frac{\Delta k l}{2}\right), \qquad (3.79)$$

where l is the propagation length through the material. Substitute

$$\text{sinc}^2 \left(\frac{\Delta k l}{2}\right) = \sin^2(\Delta k l/2)/(\Delta k l/2)^2 \qquad (3.80)$$

into Equation 3.79, we see that I_2 is proportional to $\sin^2(\Delta k l/2)$ as shown in Figure 3.16.

We are interested in measuring χ^2 under the undepleted approximation. Experimentally, we measure the varying intensity of the second harmonic wave as varying l by rotating a quartz sample with thickness d, as illustrated in Figure 3.17. The intensity variation with respect to the refractive angle (of the fundamental wave)ϕ can be obtained by substituting $l = d/\cos\phi$ into Equation 3.79 as graphically shown in Figure 3.18. The reflection coefficient also depends on ϕ in such a way that the transmittance is maximum at normal incident and decreases as $|\phi|$ increases. As a consequence, the intensity of second harmonic signals will depend on the orientation of the quartz sample as shown in Figure 3.19, which was experimentally observed by Maker et al. in 1962.

As we saw from the results of the coupled equations, in order to obtain a stable output of second harmonic generation, phase-matching condition

$$\Delta k = 2k_1 - k_2 = 0 \qquad (3.81)$$

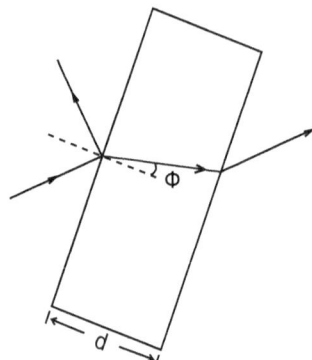

Figure 3.17: Rotating a quartz sample changes the effective thickness of the sample. Consequently, the intensity of the second harmonic generation varies with respect to the orientation of the quartz sample.

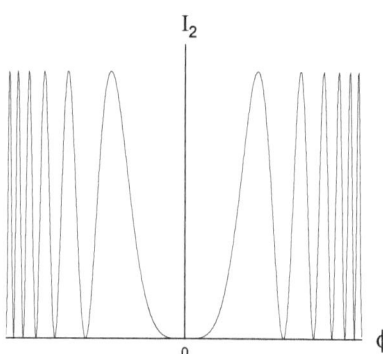

Figure 3.18: The intensity of second harmonic wave depends on the refractive angle ϕ, of the fundamental wave. Here we neglect the ϕ dependence of the transmittance.

Lecture Notes in Nonlinear Optics

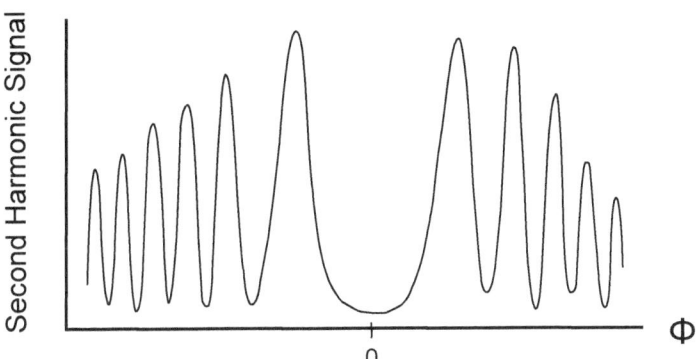

Figure 3.19: The intensity of second harmonic generation varies with respect to the orientation of a quartz sample.

is required. Substituting $k_1 = \omega n(\omega)/c$ and $k_2 = 2\omega n(2\omega)/c$ into Equation 3.81, phase-matching condition for second harmonic generation becomes

$$n(\omega) = n(2\omega). \tag{3.82}$$

Equation 3.82 suggests a possible way of achieving phase-matching by considering the dispersion of the refractive index. In the region of normal dispersion, where the refractive index monotonically increases with increasing frequency (or, decreasing wavelength), as shown in Figure 3.20. In this region, we see clearly that phase-matching condition can not be satisfied. Nevertheless, in the region of anomalous dispersion, which is in the vicinity of a linear absorption peak, the refractive index does not have a monotonic relation with frequency (or wavelength) as illustrated in Figure 3.21. Thus, it is possible to have a specific pair of frequencies ω and 2ω (or wavelengths λ and $\lambda/2$) satisfying the phase-matching condition. However, the phase-matched case is very limited. For instance, if we find an arbitrary frequency ω', the corresponding $2\omega'$ may not have the same refractive index. Experimentally, the use of anomalous dispersion is not a good way to look for phase-matching because if we change the frequency in the neighborhood of ω, then the second harmonic wave may no longer be phase-matched.

The best way to achieve phase-matching condition is by using birefringent materials. Birefringence is characterized by an index ellipsoid. Here our focus would be on uniaxial birefringence characterized by the index ellipsoid in Figure 3.22. The components of the refractive index in xy plane is independent of the polarization in that plane $n_x = n_y = n_o$, and $n_z = n_e \neq n_o$, where n_o and n_e are called ordinary and extraordinary components of the refractive index, respectively.

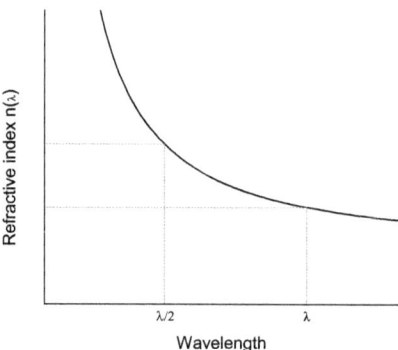

Figure 3.20: Phase-matching can not be found in the normal dispersion region.

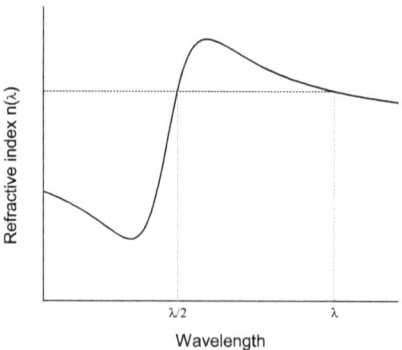

Figure 3.21: A specific phase-matched pair of wavelengths may be found in the anomalous region, but not arbitrary wavelengths satisfy the phase-matching condition.

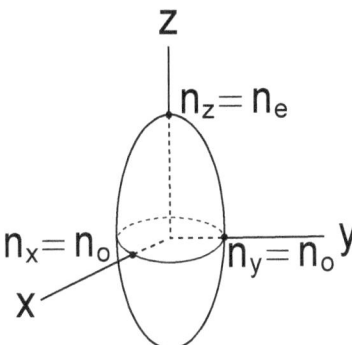

Figure 3.22: Uniaxial birefringence is characterized by an ellipsoid showing the ordinary refractive index n_o and the extraordinary refractive index n_e.

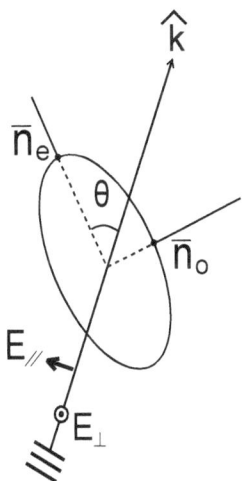

Figure 3.23: A plane wave travels along $\hat{k}$ direction. n_o is the component of the refractive index along the perpendicular polarization $E_\perp$, and n_e is the component along the parallel polarization $E_\parallel$.

A plane wave propagating in a cross section of the index ellipsoid, for example yz plane, is shown in Figure 3.23. Notice that choosing yz plane is just for the purpose of convenience. In other words, we did not lose any generality since the refractive index does not change if we rotate the index ellipsoid along z axis. Define the perpendicular polarization of the plane wave, which corresponds to n_o, to be the component of the polarization perpendicular to both z axis and the wave vector. The other component of the polarization that is perpendicular to both the perpendicular polarization and the wave vector is defined as the parallel polarization which corresponds to n_e. The refractive index will depend on the angle θ between z axis and the wave vector. Since n_o does not depend on the perpendicular polarization, which implies independent of θ, the ordinary refractive index is the component along the direction of the perpendicular polarization and can be defined as

$$n_o(\theta) = \bar{n}_o. \tag{3.83}$$

The associated light along the direction of the perpendicular polarization is called ordinary ray. The extraordinary refractive index is the component along the direction of the parallel polarization and is defined as

$$\frac{1}{n_e^2(\theta)} = \frac{\cos^2\theta}{\bar{n}_o^2} + \frac{\sin^2\theta}{\bar{n}_e^2}. \tag{3.84}$$

The associated light along the direction of the parallel polarization is called extraordinary ray. Here $\bar{n}_o$ and $\bar{n}_e$ are the principle values measured as the plane wave traveling along the semi-major axis and semi-minor axis, respectively. Evidently we have $n_e(0) = \bar{n}_o = n_o(\theta)$ and $n_e(\pi/2) = \bar{n}_e$.

As we have seen earlier in this section, the refractive index depends on wavelength, therefore this dependence should be included in the calculations. Equation 3.83 becomes

$$n_o(\theta) = \bar{n}_o(\lambda), \tag{3.85}$$

and Equation 3.84 becomes

$$\frac{1}{n_e^2(\theta)} = \frac{\cos^2\theta}{\bar{n}_o^2(\lambda)} + \frac{\sin^2\theta}{\bar{n}_e^2(\lambda)}. \tag{3.86}$$

It should be noted that to obtain phase-matching condition, the incident light must remain linearly polarized. In order to do so, the plane wave must travel in such a way that the polarization is along a semi-major axis or on a plane as shown in Figure 3.23. In contrast, elliptically polarized light can not generate second harmonic light continuously because it doesn't always satisfy phase-matching condition.

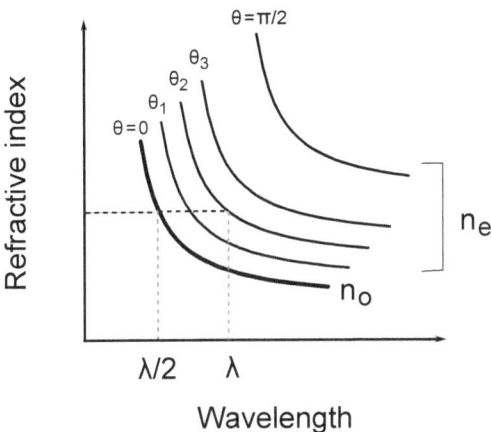

Figure 3.24: The dispersion of the refractive index under different orientations of a positive uniaxial crystal. Second harmonic generation can be achieved by varying the orientation of the crystal.

If we plot the dispersion of refractive index with respect to wavelength, since n_o does not depend on θ, there is only one dispersion-curve; but n_e depends on θ so varying θ will give us a series of curves of n_e. For a positive uniaxial crystal ($n_o < n_e$), n_e increases as θ increases (Figure 3.24), while for a negative uniaxial crystal ($n_o > n_e$), n_e decreases as θ increases (Figure 3.25). Experimentally, we continuously rotate the crystal until the orientation of the index ellipsoid gives the same refractive index for ordinary ray and extraordinary ray. Moreover, using birefringent materials also allows us to obtain different frequencies of second harmonic generation by varying the frequency (or wavelength) of the fundamental beam. In other words, the second harmonic generation is not limited at a specific wavelength.

An extraordinary ray of frequency 2ω can be generated from two incident ordinary rays of frequency ω, and this process is called type I phase-matching, which corresponds to $\chi^{(2)}_{ijj}$ with i and j referring to the extraordinary and two ordinary rays, respectively. Type II phase-matching is to generate an extraordinary ray of frequency 2ω from one ordinary ray and one extraordinary ray both with frequency ω. Type II phase-matching is corresponding to $\chi^{(2)}_{iij}$, where the first i refers to the outgoing extraordinary ray and the second i and j refer to the incident ordinary ray and extraordinary ray, respectively.

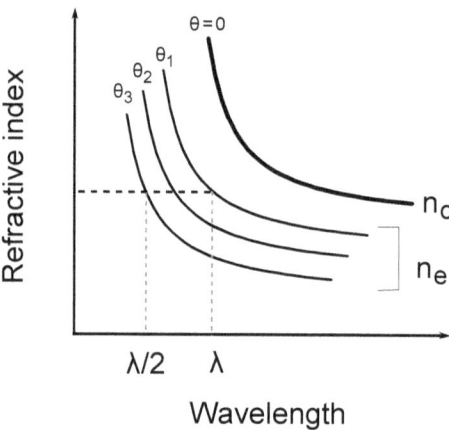

Figure 3.25: The dispersion of the refractive index under different orientations of a negative uniaxial crystal.

Chapter 4

Quantum Theory of Nonlinear Optics

4.1 A Hand-Waving Introduction to Quantum Field Theory

In this section, we motivate second quantization of the electromagnetic fields using historic arguments. Since we can calculate all the fields in free space from the vector potential, we start this section by considering $\vec{A}$.

4.1.1 Continuous Theory

Starting with Maxwell's equations one can write the wave equation in terms of $\vec{A}$. To solve the wave equation for $\vec{A}$ we choose the Coulomb gauge which is given by $\nabla \cdot \vec{A} = 0$. This condition holds only in free space where there are no free charges or currents. Given the wave equation, we get plane wave solution to $\vec{A}$ in terms of spatial and temporal coordinates given by,

$$\vec{A}(\vec{x},t) = \sum_{r=1}^{2} \sum_{\vec{k}} \vec{A}_{0r}(\vec{k}) e^{i(\vec{k}\cdot\vec{x}-\omega_k t)}. \qquad (4.1)$$

The expression is summed over all the modes of different wavevectors $\vec{k}$ and the polarizations, r, with the following transverse condition satisfied

$$\vec{k} \cdot \vec{A}_{0r}(k) = 0, \qquad (4.2)$$

for $r=1,2$.

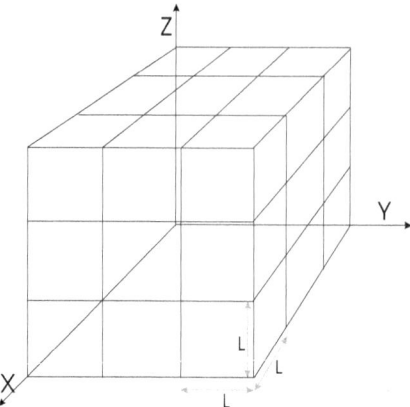

Figure 4.1: Space is equally divided into unit cells of volume L^3

The electric and magnetic fields can be calculated from $\vec{A}$ according to,

$$\vec{B} = \nabla \times \vec{A}, \tag{4.3}$$

and

$$\vec{E} = -\frac{1}{c}\frac{\partial \vec{A}}{\partial t}. \tag{4.4}$$

By substituting $\vec{A}$ from Equation 4.1 into Equations 4.3 and 4.4 we get

$$\vec{B} = \sum_{r=1}^{2}\sum_{\vec{k}} \left[\imath \vec{k} \times \vec{A}_{0r}(\vec{k})\right] \vec{k} e^{i(\vec{k}\cdot\vec{x}-\omega_k t)} \equiv \sum_{r,\vec{k}} \vec{B}_r(\vec{k}), \tag{4.5}$$

$$\vec{E} = \sum_{r=1}^{2}\sum_{\vec{k}} \frac{\imath\omega_k}{c}\vec{A}_{0r}(\vec{k}) e^{i(\vec{k}\cdot\vec{x}-\omega_k t)} \equiv \sum_{r,\vec{k}} \vec{E}_r(\vec{k}). \tag{4.6}$$

As we can see, using $k = \omega/c$ the magnitude of the fields can be written as,

$$|\vec{B}_r(\vec{k})| = |\vec{E}_r(\vec{k})| = k|\vec{A}_{0r}(\vec{k})| \equiv k|A_{0r}(\vec{k})|, \tag{4.7}$$

Since space is infinite, to avoid divergence problems in the plane wave equations, we divide the space into unit cells of dimensions L^3 as shown in Figure 4.1 and apply periodic boundary conditions to the vector potential. That is,

$$\vec{A}(x,y,z,t) = \vec{A}(x+L,y,z,t), \tag{4.8}$$

where L is an arbitrary unit length. We apply the same condition in y and z directions.

By applying the boundary conditions in Equation 4.1, we get

$$1 = e^{ik_x L}, \tag{4.9}$$
$$\Rightarrow k_x L = 2\pi n_x, \tag{4.10}$$

where n_x is an integer. By also applying the boundary conditions along y and z directions, we get

$$\vec{k} = \frac{2\pi}{L}(n_x \hat{x} + n_y \hat{y} + n_z \hat{z}). \tag{4.11}$$

So, $\vec{k}$ can only take on discrete values satisfying the boundary conditions. Eventually we can set $L \to \infty$, which will allow the choice of an arbitrary $\vec{k}$.

We can calculate the energy of a single unit cell using the corresponding Hamiltonian(from Poynting's theorem) of the radiation in a single cube.

$$H_{rad} = \frac{1}{8\pi} \int_L d^3 \vec{x} (\vec{E} \cdot \vec{E} + \vec{B} \cdot \vec{B}). \tag{4.12}$$

Here the integration is over one of the cells. The periodic boundary conditions assure that the result is independent of which cell is selected for integration.

Substituting the expressions for $\vec{E}$ and $\vec{B}$ from Equations 4.5 and 4.6 into Equation 4.12 leads to a summation over different modes. The cross terms vanish by orthonormality of the modes. For example, consider mode $\vec{k}$, which contributes an energy

$$H_{mode} = \int_0^L E_0^2 \cos^2(kx - \omega t) dx \tag{4.13}$$
$$= \int_0^L E_0^2 \left[\frac{1 + \cos(2kx - 2\omega t)}{2}\right] dx \tag{4.14}$$
$$= E_0^2 \left[\frac{1}{2} + \frac{\sin(2kx - 2\omega t)}{2k}\right]_0^L \tag{4.15}$$
$$= E_0^2 \frac{L}{2}. \tag{4.16}$$

So including the x, y and z directions, and, both the E and B fields, the Hamiltonian for a single mode can be written as,

$$H_{rad}(\vec{k}) = \frac{L^3}{16\pi}(B_0^2 + E_0^2), \tag{4.17}$$

where E_0 and B_0 are the magnitudes of $\vec{E}$ and $\vec{B}$, respectively. Or, using Equation 4.7, we can write

$$H_{rad}(\vec{k}) = \left\{\frac{L^3}{8\pi} k^2 |\vec{A}_{0r}(\vec{k})|^2\right\}, \tag{4.18}$$

which is also equal to the energy per mode.

We can write $\vec{A}_{0r} = \vec{\varepsilon}_r A_{0r}$ where A_{0r} is the magnitude of the vector potential polarized along r. To find a qualitative expression for A_{0r}, we consider the following. Lets assume there is one photon per unit cell of volume L^3. Then, according to the observation that a photon has energy $E_k = \hbar\omega_k$, we assign one photon per mode, so,

$$[Energy/mode] = \hbar\omega_k. \quad (4.19)$$

By combining Equations 4.18 and 4.19, we get

$$\hbar\omega_k = \frac{L^3}{8\pi}k^2|\vec{A}_{0r}(\vec{k})|^2 \quad (4.20)$$

$$\Rightarrow A_{0r}(\vec{k}) = \sqrt{\frac{\hbar\omega 8\pi}{L^3 k^2}}. \quad (4.21)$$

Since we only measure the real part of the quantities, we can consider only the real part of the fields or potentials. The real part of $\vec{A}$ is given by,

$$Re[\vec{A}(\vec{x},t)] = \frac{\vec{A}+\vec{A}^*}{2}. \quad (4.22)$$

From Equations 4.1, 4.20 and 4.22 we can have,

$$Re[\vec{A}(\vec{x},t)] = \sum_{\vec{k},r}\sqrt{\frac{2\pi\hbar c^2}{L^3\omega}}\vec{\varepsilon}_r\left\{a_r(\vec{k})e^{i(\vec{k}\cdot\vec{x}-\omega_k t)} + a_r^*(\vec{k})e^{-i(\vec{k}\cdot\vec{x}-\omega_k t)}\right\}. \quad (4.23)$$

Here, $a_r(\vec{k})$ and $a_r^*(\vec{k})$ are dimensionless complex quantities. Eventually we will relate them to the number of photons with wave vector $\vec{k}$.

From the Equation 4.23 we can find $\vec{E}$ using the Equation 4.6, which yields

$$\vec{E}(\vec{x},t) = i\sum_{\vec{k},r}\sqrt{\frac{2\pi\hbar\omega_k}{L^3}}\vec{\varepsilon}_r\left\{a_r(\vec{k})e^{i(\vec{k}\cdot\vec{x}-\omega_k t)} - a_r^*(\vec{k})e^{-i(\vec{k}\cdot\vec{x}-\omega_k t)}\right\}. \quad (4.24)$$

We can write a similar expression for $\vec{B}(\vec{x},t)$.

4.1.2 Second Quantization

Recalling the harmonic oscillator, we have,

$$\hat{x} = \frac{\sqrt{2\hbar m\omega}}{2m\omega}(a+a^\dagger), \tag{4.25}$$

$$\hat{p} = \frac{-i\sqrt{2\hbar m\omega}}{2}(a-a^\dagger), \tag{4.26}$$

$$H = \frac{p^2}{2m} + \frac{1}{2}m\omega^2 x^2 = \hbar\omega\left(a^\dagger a + \frac{1}{2}\right), \tag{4.27}$$

with $[a,a^\dagger] = 1$ and $[x,p] = i\hbar$.

Using the analogy with a simple harmonic oscillator(E and B behaves like x and p), we will make the assumption that $a_r(\vec{k})$ is *annihilation operator* and $a_r^\dagger(\vec{k})$ is the *creation operator*, annihilating and creating one quantum of energy with wave vector $\vec{k}$ and polarization r.

So, the corresponding commutator is given by,

$$\left[a_r(\vec{k}), a_{r'}^\dagger(\vec{k'})\right] = \delta_{k,k'}\delta_{r,r'}. \tag{4.28}$$

With this assumption made, we calculate the expression for the Hamiltonian as follows.(Note that the validity of the assumption is based on the success of the resulting theory in its predictions.)

By using the orthogonality condition,

$$\int_0^L e^{ikx} e^{ik'x} dx = \delta_{kk'} L, \tag{4.29}$$

one can prove that in calculating H_{rad}, only cross terms $(\vec{k}, -\vec{k})$ from the same mode survive. From Equations 4.18 and 4.23, H_{rad} is given by,

$$H_{rad} = \sum_{\vec{k},r} \frac{1}{8\pi}\left\{2\pi\hbar\omega_k 2\left[a_r(\vec{k})a_r(\vec{k})^\dagger + a_r(\vec{k})^\dagger a_r(\vec{k})\right]\right\}. \tag{4.30}$$

Here $|\vec{\varepsilon}_r|$ is unity. With the commuter $[a_r(\vec{k}), a_r^\dagger(\vec{k})] = 1$ we get

$$H_{rad} = \sum_{\vec{k},r} \hbar\omega_k \left[a_r^\dagger(\vec{k})a_r(\vec{k}) + \frac{1}{2}\right]. \tag{4.31}$$

As we know, $a_r(\vec{k})a_r^\dagger(\vec{k})$ is the number operator $N_r(\vec{k})$. Thus

$$H_{rad} = \sum_{\vec{k},r} \hbar\omega_k \left(N_r(\vec{k}) + \frac{1}{2}\right). \tag{4.32}$$

Here an interesting point to note is that the second term on the right hand side goes to infinity independent of the number of photons as there are an infinite number of possible modes. That is,

$$\sum_{\vec{k},r} \hbar\omega \frac{1}{2} = \infty. \tag{4.33}$$

Since we always deal with differences of energies, we ignore this term in further calculations. The Hamiltonian is thus given by,

$$H_{rad} = \sum_{r,\vec{k}} \hbar\omega_k N_r(\vec{k}). \tag{4.34}$$

4.1.3 Photon-Molecule Interactions

We now consider the Hamiltonian of an electron in an atom or molecule, or

$$H_e = \frac{\vec{p}^2}{2m} + V. \tag{4.35}$$

Here V includes all internal forces on the electron due to the nucleus and other electrons. In an external electrical field, the Hamiltonian becomes

$$H' = \frac{1}{2m}\left(\vec{p} - \frac{q\vec{A}}{c}\right)^2 + V + \phi_m \tag{4.36}$$

$$= \frac{1}{2m}\left(\vec{p}^2 + \frac{e^2|\vec{A}|^2}{c} + \frac{2e\vec{A}\cdot\vec{p}}{c}\right) + V, \tag{4.37}$$

where the charge of electron is $q = -e$. $\phi_m = 0$, $[\vec{A},\vec{p}] = 0$ in the Coulomb gauge. Here we also neglect higher order terms such as $|\vec{A}|^2$. The total Hamiltonian of the electron is thus given by,

$$H' = \frac{1}{2m}\left(\vec{p}^2 + \frac{2e\vec{A}\cdot\vec{p}}{c}\right) + V, \tag{4.38}$$

or,

$$H' = H_e + H_I \tag{4.39}$$

where the interaction Hamiltonian H_I is given by,

$$H_I = \frac{e\vec{A}\cdot\vec{p}}{mc}. \tag{4.40}$$

Lecture Notes in Nonlinear Optics

The Hamiltonian of the whole system including the fields, is given by,

$$H = H_e + H_{rad} + H_I = H_0 + H_I, \quad (4.41)$$

where eigen functions for H_e and H_{rad} are known and we treat the mixing term H_I using perturbation theory.

The potential energy of a dipole in an electric field $\vec{E}$ is given by,

$$U = -\vec{\mu} \cdot \vec{E} = e\vec{x} \cdot \vec{E}, \quad (4.42)$$

where e is the magnitude of the electron charge. Here we are ignoring higher order moments and higher-order terms in E.

Substituting expression for E from 4.24,

$$U = e\vec{x} \cdot \vec{E} \quad (4.43)$$

$$= ie \sum_{\vec{k},r} \sqrt{\frac{2\pi\hbar\omega_k}{L^3}} \vec{x} \cdot \vec{\varepsilon}_r \left[a_r(\vec{k}) e^{i(\vec{k} \cdot \vec{x} - \omega_k t)} + H.A. \right], \quad (4.44)$$

where H.A. is the hermitian adjoint due to the operators for electron position $\vec{x}$ and annihilation of a photon, a_r. One important point to note is that $\vec{x}$ is an operator in electron space whereas a_r is an operator in photon space.

For a molecule that is small compared with the lights wavelength,

$$e^{i(\vec{k} \cdot \vec{x})} = 1 + i(\vec{k} \cdot \vec{x}) + ... \quad (4.45)$$

For an atom or molecules, $\vec{x}$ is the position coordinate of an electron and its in the order of few angstroms. Then, $\vec{k} \cdot \vec{x}$ is very small for visible light. For light of wavelength 1×10^{-4}cm and atom of 1Å radius,

$$\vec{k} \cdot \vec{x} \sim \frac{2\pi}{\lambda} \sim 2\pi \times 10^{-4}. \quad (4.46)$$

Thus we assume that,

$$e^{i(\vec{k} \cdot \vec{x})} \sim 1 = constant. \quad (4.47)$$

Equation 4.47 is often called as *dipole approximation*. Some materials may not have dipole moment $\vec{\mu}$ but have an octuple moment or, in some cases, the size of the system may be comparable to the wavelength of the light. In those cases, the dipole approximation is not valid.

Since H_e and H_{rad} belong to different spaces, we can write,

$$[H_e, H_{rad}] = 0. \quad (4.48)$$

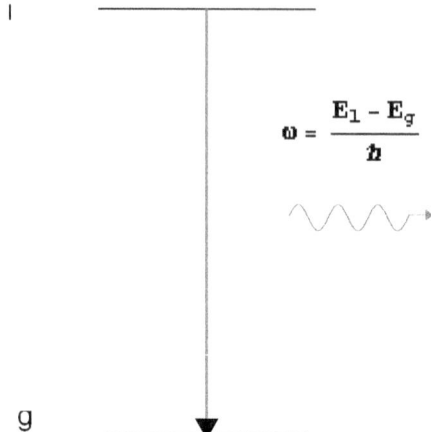

Figure 4.2: Spontaneous emission of a photon due to de-excitation

Now, construct a wave function as a direct-product of state vectors $|\psi_e\rangle$ and $|\psi_{rad}\rangle$, which are eigenfunctions of H_e and H_{rad}, respectively,

$$|\psi\rangle = |\psi_e\rangle |\psi_{rad}\rangle. \tag{4.49}$$

Then,

$$\begin{aligned} H_0 |\psi_e\rangle |\psi_{rad}\rangle &= (H_e |\psi_e\rangle) |\psi_{rad}\rangle + |\psi_e\rangle H_{rad} |\psi_{rad}\rangle && (4.50) \\ &= (E_{rad} + E_e) |\psi\rangle, && (4.51) \end{aligned}$$

where E_{rad}, E_e are eigen values of the corresponding state vectors.

We define the state vectors as follows,

$$|\psi\rangle = |n; N_{\vec{k}_1, r_1}, N_{\vec{k}_2, r_2}, N_{\vec{k}_3, r_3} ...\rangle, \tag{4.52}$$

where n represents quantum number(s) of electrons(k,l,m etc.,) and $N_{\vec{k}_1, r_1}$ represents the number of photons with wavevector $\vec{k} = \vec{k}_1$ and polarization r_1, and so on.

Consider a two level system with ground state(g) and excited state(l). If we assume the electron is initially in the excited state space(Figure 4.2) then the probability of deexcitation is proportional to the transition moment. That is,

$$\begin{aligned} \wp_{fi} &\propto |\langle \psi_f | H_I | \psi_i \rangle|^2 && (4.53) \\ &= \kappa |\langle \psi_f | \frac{e\vec{A} \cdot \vec{p}}{mc} | \psi_i \rangle|^2, && (4.54) \end{aligned}$$

Lecture Notes in Nonlinear Optics 93

where κ is a proportionality constant. Here, by energy conservation, $E_l - E_g = \hbar\omega$. The wave functions for the states of the system are given by,

$$|\psi_i\rangle = |l; 0, 0, 0...\rangle, \tag{4.55}$$

$$|\psi_f\rangle = |g; 0, 0, 1_{k=\frac{\omega}{c}, r}, 0, 0...\rangle, \tag{4.56}$$

where the one photon in $|\psi_f\rangle$ corresponds to the photon emitted due to de-excitation.

4.1.4 Transition amplitude in terms of $\vec{E}$

We saw that the probability for making a transition between states due to the absorption or emission of light is related to matrix elements of $\vec{A} \cdot \vec{p}$. Here we show that this is equivalent to using matrix elements of $\vec{\mu} \cdot \vec{E}$.

By considering the dipole approximation discussed earlier, we can write Equations 4.53 and 4.54 as follows,

$$\wp_{fi} = \kappa \frac{e^2}{m^2 c^2} |\langle 1_{k=\frac{\omega}{c}, r} | \vec{A} | 0 \rangle \cdot \langle g | \vec{p} | l \rangle|^2. \tag{4.57}$$

Now to calculate $\langle g | \vec{p} | l \rangle$, consider the commutator

$$[\vec{x}, H_e] = i\hbar \frac{\vec{p}}{m}. \tag{4.58}$$

Matrix elements of the above commutator in electron space are given by,

$$\langle g | \vec{x} H_e - H_e \vec{x} | l \rangle = \frac{i\hbar}{m} \langle g | \vec{p} | l \rangle, \tag{4.59}$$

$$(E_l - E_g)\langle g | \vec{x} | l \rangle = \frac{i\hbar}{m} \langle g | \vec{p} | l \rangle, \tag{4.60}$$

$$\Rightarrow \langle g | \vec{p} | l \rangle = \frac{m\omega}{i} \langle g | \vec{x} | l \rangle. \tag{4.61}$$

When we have used the fact that $E_l - E_g = \hbar\omega$, that is emitted photon energy is the energy difference between the two states.

Substituting Equation 4.61 into Equation 4.57 we get,

$$\wp_{fi} = \kappa \frac{e^2}{m^2 c^2} |\langle 1_{k=\frac{\omega}{c}, r} | \vec{A} | 0 \rangle \cdot \left(\frac{m\omega}{i}\right) \langle g | \vec{x} | l \rangle|^2 \tag{4.62}$$

$$= \kappa \frac{-\omega^2}{c^2} |\langle 1_{k=\frac{\omega}{c}, r} | \vec{A} | 0 \rangle \cdot \langle g | e\vec{x} | l \rangle|^2. \tag{4.63}$$

Since there is only one photon created, we can write the vector potential as

$$\vec{A} = a^\dagger A_0 \vec{\varepsilon}_r, \tag{4.64}$$

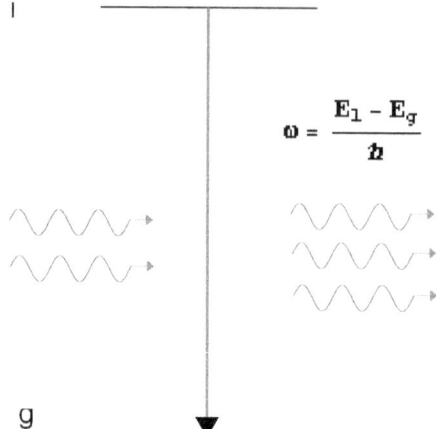

Figure 4.3: Stimulated emission from deexcited system in the presence of identical photons

so, the transition probability is given by,

$$\wp_{fi} = \kappa A_0^2 \frac{-\omega^2}{c^2} |\langle 1_{k=\frac{\omega}{c},r} | \vec{a}^\dagger | 0 \rangle \langle g | e\vec{x} \cdot \vec{\varepsilon}_r | l \rangle|^2. \tag{4.65}$$

Since $\frac{\omega^2}{c^2} A_0^2$ has same units as of E^2, we can write

$$\wp_{fi} = \kappa |\langle g | e\vec{x} \cdot \vec{E} | l \rangle|^2 \tag{4.66}$$
$$= \kappa |\langle g | \mu \cdot \vec{E} | l \rangle|^2. \tag{4.67}$$

Thus, the interaction hamiltonian can equally expressed as $\mu \cdot \vec{E}$ in the dipole approximation and Equation 4.67 is equivalent to Equation 4.54. We can use whatever form be simpler for the particular application.

4.1.5 Stimulated Emission

Consider a two level system in its excited state. Initially there are N photons with wave vector $\vec{k}$ and polarization direction r as shown in Figure 4.3. When the system de-excites, a photon will be emitted, so if the de-excitation is driven by stimulated emission, the final state consists of $N + 1$ photons with the same wave vector and polarization. The probability of a transition is proportional to the matrix element of the interaction hamiltonian. Initially,

$$N_{\vec{k}r} = N. \tag{4.68}$$

Lecture Notes in Nonlinear Optics 95

The final and initial wavefunctions are,

$$|\psi_i\rangle = |i; N\rangle, \quad (4.69)$$
$$|\psi_f\rangle = |f; N+1\rangle, \quad (4.70)$$

Here we assume all other occupation numbers are 0. Then the transition strength is given by

$$\wp_{fi} \propto |\langle \psi_f | H_I | \psi_i \rangle|^2 \quad (4.71)$$
$$= |\langle f | e\vec{x} | i \rangle \cdot \langle N+1 | \vec{E} | N \rangle|^2 \quad (4.72)$$
$$= |\langle f | \vec{\mu} | i \rangle|^2 |\langle N+1 | a_r^\dagger(\vec{k}) | N \rangle|^2 \quad (4.73)$$
$$= |\vec{\mu}_{fi}|^2 \left(\sqrt{N+1}\right)^2 \quad (4.74)$$
$$= |\vec{\mu}_{fi}|^2 (N+1). \quad (4.75)$$

Thus, if $N = 0$, the result is spontaneous emission. When N is large, the probability of de-excitation increases with the number of photons already present, which is the process of stimulated emission.

Problem 4.1-1: A physicist discovers a process in which light at frequency ω_1 and ω_2 is converted by a sample to light at frequency $2\omega_1 + \omega_2$ only when all the light is polarized along $\hat{x}$. The intensity of the output peaks at frequency ω_0 and has the form $\exp\left(-[\{(2\omega_1+\omega_2)-\omega_0\}/\omega_0]^2\right)$. Express the *simplest* interaction Hamiltonian H_I in terms of photon creation and annihilation operators that describes this process. The constant of proportionality is unimportant.

4.2 Time-Dependent Perturbation Theory

Our goal is to calculate the nonlinear susceptibility of a quantum system. As is always the case, the first step requires a calculation of the induced dipole moment as a function of the electric field, which is then differentiated according to Equation 1.35 to get the nonlinear susceptibilities.

All of the information about a quantum system is contained in the wavefunction and any particular observable is simply the expectation value of an operator that represents the observable. Thus, we need to calculate the wavefunction of a molecule in an electric field, from which we calculate the expectation value of the dipole moment, which we differentiate with respect to the electric field to get the susceptibility.

We therefore define $|\psi_n(\vec{E})\rangle$ as the nth energy eigenstate of an atom or molecule in the presence of electric field $\vec{E}(t)$, where

$$\vec{E}(t) = \sum_{i=1}^{\text{\# of fields}} \vec{E}^{\omega_i}(t). \tag{4.76}$$

At zero temperature the atom/molecule is in its ground state, so the polarization is given by

$$\vec{P}(\vec{E}) = \langle \psi_0(\vec{E}) | \vec{P} | \psi_0(\vec{E}) \rangle, \tag{4.77}$$

where $|\psi_0(\vec{E})\rangle$ is the perturbed ground state. The susceptibility is then given by,

$$\chi^{(n)}_{ijk...}(-\omega_\sigma;\omega_1,\omega_2,...) = \frac{\partial^n}{\partial E^{\omega_1}_j \partial E^{\omega_2}_k ...} \langle \psi_0(\vec{E}) | P_i | \psi_0(\vec{E}) \rangle \frac{1}{D'} \bigg|_{\vec{E}=0} \tag{4.78}$$

where D' is the frequency dependent degeneracy denominator introduced earlier in Chapter 2.

4.2.1 Time-Dependent Perturbation Theory

Our strategy is to determine the wavefunction of the molecule in the presence of the electric field using the zero-field wavefunctions as a basis. The field dependence will be calculated using dipole coupling between the field and the molecule as a time-dependent perturbation.

Let H_0 be the molecular Hamiltonian without perturbation, that is, when the electric field is not present, the time evolution of a state $|\psi\rangle$ is given by

$$i\hbar \frac{\partial}{\partial t} |\psi\rangle = H_0 |\psi\rangle. \tag{4.79}$$

We consider as a perturbation

$$V(t) = -\vec{\mu} \cdot \vec{E}(t), \tag{4.80}$$

which is due to the interaction energy of the electric dipole moment of the system and the applied electric field, where

$$\vec{E}(t) = \frac{1}{2} \sum_p \vec{E}(\omega_p) e^{-i\omega_p t} \quad (\omega_{-p} = -\omega_p), \tag{4.81}$$

where $\vec{\mu}$ is the dipole moment of the atom/molecule.

We define the unperturbed eigenstates as

$$\left|\psi_n^{(0)}(t)\right\rangle = \left|u_n^{(0)}\right\rangle e^{-i\omega_n t}, \tag{4.82}$$

where $\omega_n = E_n/\hbar$ and $\left|u_n^{(0)}\right\rangle$ are the eigenstates of the original Hamiltonian. We can then write the total Hamiltonian, H, as the sum of the molecular Hamiltonian and the perturbation multiplied by a small parameter λ,

$$H = H_0 + \lambda V. \tag{4.83}$$

The perturbed states can be expressed as a sum of different orders of correction that are labeled by λ^m,

$$\left|\psi_n(t)\right\rangle = \sum_{m=0}^{\infty} \lambda^m \left|\psi_n^{(m)}(t)\right\rangle, \tag{4.84}$$

where m is the order of correction to $\left|\psi\right\rangle$.

Substituting Equation 4.84 into the Schrodinger equation, and keeping all terms of order m, we get

$$i\hbar \frac{\partial}{\partial t} \left|\psi_n^{(m)}(t)\right\rangle = H_0 \left|\psi_n^{(m)}(t)\right\rangle + V(t) \left|\psi_n^{(m-1)}(t)\right\rangle. \tag{4.85}$$

Using the fact that the eigenvectors, $\left|\psi_n^{(m)}(t)\right\rangle$, can be expanded in terms of unperturbed states $\left|\psi_l^{(0)}(t)\right\rangle$ with coefficients $a_{ln}^{(m)}(t)$,

$$\left|\psi_n^{(m)}(t)\right\rangle = \sum_l a_{ln}^{(m)}(t) \left|\psi_l^{(0)}(t)\right\rangle \tag{4.86}$$

the ground state wavefunction in the presence of an electric field can be expressed as

$$\left|\psi_0^{(m)}(t)\right\rangle = \sum_l a_l^{(m)}(t) \left|\psi_l^{(0)}(t)\right\rangle, \tag{4.87}$$

where $a_l^{(m)}(t) \equiv a_{l0}^{(m)}(t)$. Substituting Equation 4.87 into Equation 4.85 we get

$$i\hbar \left(\sum_l \dot{a}_l^{(m)} \left|\psi_l^{(0)}(t)\right\rangle + \sum_l a_l^{(m)}(-i\omega_l) \left|\psi_l^{(0)}(t)\right\rangle \right)$$
$$= \sum_l E_l a_l^{(m)} \left|\psi_l^{(0)}(t)\right\rangle + V(t) \sum_l a_l^{(m-1)} \left|\psi_l^{(0)}(t)\right\rangle. \tag{4.88}$$

Operating on Equation 4.88 from left with $\langle u_p^{(0)}|$, we get

$$i\hbar \dot{a}_p^{(m)} e^{-i\omega_p t} + \hbar \omega_p a_p^{(m)} e^{-i\omega_p t} = E_p a_p^{(m)} e^{-i\omega_p t}$$
$$+ \sum_l \langle u_p^{(0)}|V(t)|u_l^{(0)}\rangle a_l^{(m-1)} e^{-i\omega_l t}. \quad (4.89)$$

By defining ω_{lp} and $V_{pl}(t)$ as

$$\omega_{lp} = \omega_l - \omega_p$$
$$V_{pl}(t) = \langle u_p^{(0)}|V(t)|u_l^{(0)}\rangle \quad (4.90)$$

we can write $\dot{a}_p^{(m)}$ as

$$\dot{a}_p^{(m)} = \frac{1}{i\hbar} \sum_l V_{pl}(t) a_l^{(m-1)}(t) e^{-i\omega_{lp} t}. \quad (4.91)$$

Integration of Equation 4.91 gives

$$a_p^{(m)}(t) = \frac{1}{i\hbar} \sum_l \int_{-\infty}^{t} V_{pl}(t) a_l^{(m-1)}(t) e^{-i\omega_{lp} t} dt. \quad (4.92)$$

Since initially the system is in its ground state, $a_l^{(0)} = \delta_{l,0}$, Equation 4.92 with the help of Equation 4.80 and Equation 4.81 yields

$$a_p^{(1)}(t) = \frac{1}{i\hbar} \sum_l \int_{-\infty}^{t} \vec{\mu}_{pl} \cdot \frac{1}{2} \sum_q \vec{E}(\omega_q) e^{-i\omega_q t} \delta_{l,0} e^{-i\omega_{lp} t} dt. \quad (4.93)$$

The integral at negative infinity vanishes if we make Ω_{p0} slightly complex, so Equation 4.93 yields

$$a_p^{(1)}(t) = \frac{1}{2\hbar} \sum_q \frac{\vec{\mu}_{p0} \cdot \vec{E}(\omega_q)}{\Omega_{p0} - \omega_q} \exp\left[i(\Omega_{p0} - \omega_q)t\right]. \quad (4.94)$$

The coefficient $a_p^{(2)}$ is derived by substituting Equation 4.94 into Equation 4.92, and gives

$$a_r^{(2)}(t) = \frac{1}{4\hbar^2} \sum_{q,s} \sum_v \frac{\left[\vec{\mu}_{rv} \cdot \vec{E}(\omega_s)\right]\left[\vec{\mu}_{v0} \cdot \vec{E}(\omega_q)\right]}{(\Omega_{r0} - \omega_q - \omega_s)(\Omega_{v0} - \omega_q)} \exp\left[i(\Omega_{r0} - \omega_q - \omega_s)t\right], \quad (4.95)$$

Lecture Notes in Nonlinear Optics 99

and $a_p^{(3)}$ is calculated using Equation 4.95 in Equation 4.92,

$$\begin{aligned}
a_p^{(3)}(t) &= \frac{1}{8\hbar^3} \sum_{q,r,s} \sum_{m,n} \frac{\left[\vec{\mu}_{pm} \cdot \vec{E}(\omega_s)\right]\left[\vec{\mu}_{mn} \cdot \vec{E}(\omega_q)\right]\left[\vec{\mu}_{n0} \cdot \vec{E}(\omega_r)\right]}{(\Omega_{p0} - \omega_q - \omega_r - \omega_s)(\Omega_{m0} - \omega_q - \omega_r)(\Omega_{n0} - \omega_q)} \\
&\times \exp\left[i\left(\Omega_{p0} - \omega_q - \omega_r - \omega_s\right)t\right].
\end{aligned} \tag{4.96}$$

Recall that $\Omega_{nm} = \Omega_n - \Omega_m$ is the Bohr frequency between states n and m.

We assume that the system remains in the ground state in the presence of the field; but, because the field is time-dependent, the ground state wavefunction will evolve with time and is of the form,

$$|\psi_0(t)\rangle = |0\rangle + \lambda \sum_p a_p^{(1)}(t) |p\rangle e^{-i\Omega_{p0}t} + \lambda^2 \sum_p a_p^{(2)}(t) |p\rangle e^{-i\Omega_{p0}t} + \ldots, \tag{4.97}$$

where we have used the fact that

$$\left|\psi_p^{(0)}(t)\right\rangle = \left|u_p^{(0)}\right\rangle e^{-i\Omega_{p0}t}. \tag{4.98}$$

Note that since only energy differences are important, Ω_p and Ω_{p0} are equivalent.

Using Equation 4.97, the expectation value of the dipole moment will be of the form

$$\begin{aligned}
\langle\vec{\mu}\rangle(t) &= \left(\langle 0| + \lambda \sum_p a_p^{(1)*}(t) \langle p| e^{+i\Omega_{p0}^* t} + \ldots\right) \vec{\mu} \\
&\times \left(|0\rangle + \lambda \sum_p a_p^{(1)}(t) |p\rangle e^{-i\Omega_{p0}t} + \lambda^2 \sum_p a_p^{(2)}(t) |p\rangle e^{-i\Omega_{p0}t} + \ldots\right).
\end{aligned} \tag{4.99}$$

Here, we allow Ω_{p0} to be complex. The meaning of this will be clarified later.

Given Equation 4.99, the nth-order nonlinear susceptibilities will be related to dipole moment to order λ^n.

4.2.2 First-Order Susceptibility

Using Equation 4.99, the dipole moment to first order in λ is given by

$$\langle\vec{\mu}\rangle^{(1)}(t) = \sum_p a_p^{(1)*}(t) \langle p| e^{+i\Omega_{p0}^* t} \vec{\mu} |0\rangle + \langle 0| \vec{\mu} \sum_p a_p^{(1)}(t) |p\rangle e^{-i\Omega_{p0}t}. \tag{4.100}$$

We evaluate Equation 4.100 using Equation 4.94. However, the calculation is made simpler if we first recast $a_p^{(1)*}$, as follows.

Taking the complex conjugate of Equation 4.94, we get,

$$a_p^{(1)*}(t) = \frac{1}{2\hbar} \sum_q \frac{\vec{\mu}_{0p} \cdot \vec{E}(-\omega_q)}{\Omega_{p0}^* - \omega_q} \exp\left[-i\left(\Omega_{p0}^* - \omega_q\right)t\right], \qquad (4.101)$$

where we have used $\vec{\mu}_{p0}^* = \vec{\mu}_{0p}$. But, recall that q sums over positive and negative frequencies with $\omega_{-q} = -\omega_q$. Thus, if we invert the sum by taking $q \to -q$, the sum remains unchanged since we are still summing over all q. This re-indexing of Equation 4.101 leads to,

$$a_p^{(1)*}(t) = \frac{1}{2\hbar} \sum_q \frac{\vec{\mu}_{0p} \cdot \vec{E}(\omega_q)}{\Omega_{p0}^* + \omega_q} \exp\left[-i\left(\Omega_{p0}^* + \omega_q\right)t\right]. \qquad (4.102)$$

Substituting Equations 4.102 and 4.94 into Equation 4.100 gives,

$$\langle \vec{\mu} \rangle^{(1)}(t) = \frac{1}{2\hbar} \sum_p \sum_q \left(\frac{\vec{\mu}_{0p} \cdot \vec{E}(\omega_q)}{\Omega_{p0}^* + \omega_q} \vec{\mu}_{p0} + \vec{\mu}_{0p} \frac{\vec{\mu}_{p0} \cdot \vec{E}(\omega_q)}{\Omega_{p0} - \omega_q} \right) e^{-i\omega_q t}, \qquad (4.103)$$

where we have used the shorthand notation $\vec{\mu}_{nm} = \langle n|\vec{\mu}|m\rangle$. But the polarization is given by,

$$\vec{P}(t) = \sum_\omega \left(\frac{\vec{P}(\omega)}{2} e^{-i\omega t} + \frac{\vec{P}(-\omega)}{2} e^{i\omega t} \right) = N \langle \vec{\mu} \rangle^{(1)}(t), \qquad (4.104)$$

so projecting out the ω fourier component yields,

$$\vec{P}(\omega_\sigma) = \frac{N}{\hbar} \sum_p \left(\frac{\vec{\mu}_{0p} \cdot \vec{E}(\omega_\sigma)}{\Omega_{p0}^* + \omega_q} \vec{\mu}_{p0} + \vec{\mu}_{0p} \frac{\vec{\mu}_{p0} \cdot \vec{E}(\omega_\sigma)}{\Omega_{p0} - \omega_\sigma} \right). \qquad (4.105)$$

The first-order susceptibility is calculated using Equation 4.78 and yields,

$$\chi_{ij}^{(1)}(-\omega_\sigma;\omega_\sigma) = -N \sum_p \left(\frac{\vec{\mu}_{0p}^j \vec{\mu}_{p0}^i}{\mathcal{E}_{p0}^* + \hbar\omega_\sigma} + \frac{\vec{\mu}_{0p}^i \vec{\mu}_{p0}^j}{\mathcal{E}_{p0} - \hbar\omega_\sigma} \right), \qquad (4.106)$$

where $\vec{\mu}_{0p}^k$ is the kth cartesian component of $\vec{\mu}_{0p}$, $\mathcal{E}_{p0} = \hbar\Omega_{p0}$, and where we have used the fact that the definition of $\chi_{ij}^{(1)}$ assigns i to the direction of the polarization $\vec{P}(\omega_\sigma)$, and j to the direction of the applied field $\vec{E}(\omega_\sigma)$.

4.2.3 Nonlinear Susceptibilities and Permutation Symmetry

The second- and third-order (and higher order) susceptibilities can be calculated using a similar approach as the first-order one by first finding $\langle \vec{\mu} \rangle^{(n)}(t)$ from $\langle \vec{\mu} \rangle(t)$ using n^{th}-order λ terms and differentiating the result according to Equation 4.78 after selecting the ω_σ Fourier component. The second-order susceptibility is then given by,

$$\chi^{(2)}_{ijk}(-\omega_\sigma;\omega_1,\omega_2) = \frac{Ne^3}{\hbar^2}\mathcal{P}_\mathcal{I}\sum_{m,n}\left[\frac{r^i_{0n}r^j_{nm}r^k_{m0}}{(\Omega_{n0}-\omega_1-\omega_2)(\Omega_{m0}-\omega_2)}\right.$$
$$+ \frac{r^j_{0n}r^i_{nm}r^k_{m0}}{(\Omega_{n0}+\omega_1)(\Omega_{m0}-\omega_2)}$$
$$+ \left.\frac{r^j_{0n}r^k_{nm}r^i_{m0}}{(\Omega_{n0}+\omega_1)(\Omega_{m0}+\omega_2+\omega_1)}\right], \tag{4.107}$$

where we have expressed $\vec{\mu}$ as $e\vec{r}$, and $\mathcal{P}_\mathcal{I}$ is the 'intrinsic permutation operator'. It dictates that we take an average over all permutations of ω_1 and ω_2 with simultaneous permutations of the Cartesian components. For example, the first term in brackets in Equation 4.107 under permutation yields

$$\frac{1}{2}\left[\frac{r^i_{0n}r^j_{nm}r^k_{m0}}{(\Omega_{n0}-\omega_1-\omega_2)(\Omega_{m0}-\omega_2)} + \frac{r^i_{0n}r^k_{nm}r^j_{m0}}{(\Omega_{n0}-\omega_2-\omega_1)(\Omega_{m0}-\omega_1)}\right]. \tag{4.108}$$

Equation 4.107 can be rewritten using the full permutation operator $\mathcal{P}_\mathcal{F}$. When ω is far from resonance ω_σ can be permuted with ω_n with simultaneous permutation of the Cartesian components, so the first order susceptibility can be rewritten as,

$$\chi^{(1)}_{ij}(-\omega_\sigma;\omega_\sigma) = \frac{Ne^2}{\hbar}\mathcal{P}_\mathcal{F}\sum_m\left(\frac{r^i_{0m}r^j_{m0}}{(\Omega_{m0}-\omega_\sigma)}\right). \tag{4.109}$$

Next we consider the second-order susceptibility. After some manipulations, the ground state is found to be excluded from the sum and the matrix elements of r have subtracted from them r_{00}, thus the second order susceptibility is given by,

$$\chi^{(2)}_{ijk}(-\omega_\sigma;\omega_1,\omega_2) = \frac{Ne^3}{2\hbar^2}\mathcal{P}_\mathcal{F}\sum_{m,n}{}'\left(\frac{r^i_{0n}\bar{r}^j_{nm}r^k_{m0}}{(\Omega_{m0}-\omega_\sigma)(\Omega_{n0}-\omega_2)}\right), \tag{4.110}$$

where the prime indicates that the sum excludes the ground state, and $\bar{r}^j = r^j - r_{00}^j$. Equation 4.107 for $\chi_{ijk}^{(2)}$ indeed reduces to Equation 4.110 upon the full permutation operation.

And similarly the third order susceptibility is given by,

$$\chi_{ijkh}^{(3)}(-\omega_\sigma;\omega_1,\omega_2,\omega_3) = \frac{Ne^4}{6\hbar^3}\mathscr{P}_{\mathscr{F}}\sum_{m,n,l}{}'$$

$$\frac{\bar{r}_{0l}^i\bar{r}_{ln}^j r_{nm}^k r_{m0}^h}{(\Omega_{l0}-\omega_\sigma)(\Omega_{n0}-\omega_2-\omega_3)(\Omega_{m0}-\omega_3)}.$$

(4.111)

Problem 4.2-1: Using time dependent perturbation theory, we found that

$$\dot{a}_p^{(m)}(t) = \frac{1}{i\hbar}\sum_l \langle u_p^{(0)}|V(t)|u_l^{(0)}\rangle a_l^{(m-1)}(t)e^{-i\omega_{lp}t}, \qquad (4.112)$$

where $V(t) = e\vec{r}\cdot\vec{E}$. Express the electric field in terms of photon creation and annihilation operators and solve for $a_r^{(m)}(t)$ using the approximation $\vec{k}\cdot r = 0$.

4.3 Using Feynman-Like Diagrams

4.3.1 Introduction

A **Feynman diagram** is a pictorial technique to evaluate the nonlinear susceptibilities, and it is a simple way to picture the corresponding physical process when photons are being absorbed or emitted. A Feynman diagram can be used to evaluate a nonlinear process to various orders of perturbation, and allows one to write down immediately the mathematical expression associated with that particular process. Thus it is a convenient tool for evaluating complicated higher order nonlinear optical responses.

Consider the two collision processes shown in Figure 4.4. Figure 4.4 shows the physical processes in space, and Figure 4.5 shows the same events on a space-time diagram that can be diagrammatically evaluated.

Figure 4.6 shows a Feynman diagram for the electromagnetic interaction between two charged particles such as electrons. Since we understand that the vertical axis is time, and horizontal axis is space, we will omit the axes in the diagrams:

Lecture Notes in Nonlinear Optics   103

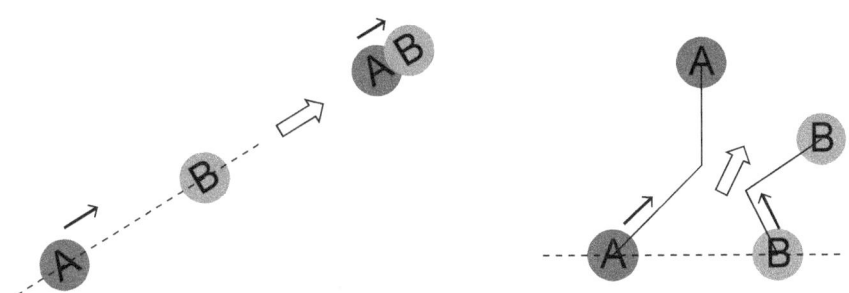

Figure 4.4: Particle trajectories of collision. (left): Inelastic collision; (right): Elastic collision.

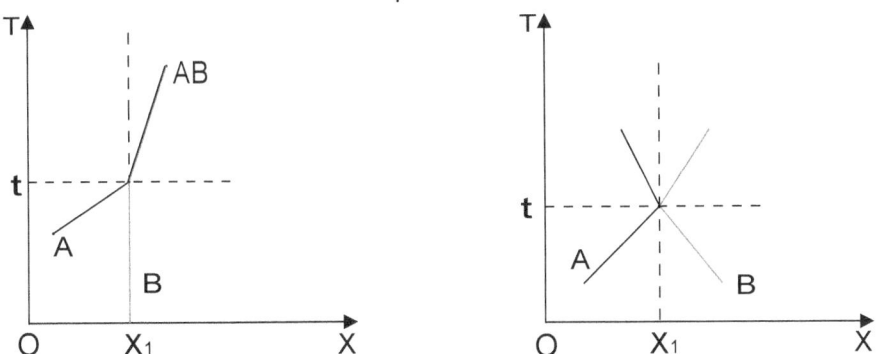

Figure 4.5: Space-time diagram for above.

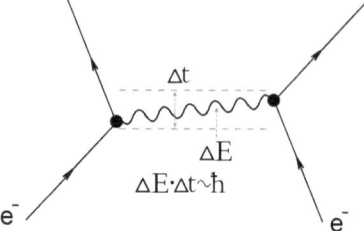

Figure 4.6: The Feynman diagram that represents the interaction between two electrons.

A incident electron emits a photon for no apparent reason and then deflects. This is a virtual process because energy is not conserved. A second incident electron absorbs the photon emitted by the first electron, and also deflects. In this diagram, the electron moves more slowly than the photon; the electron on the left gives off energy, so its slope decreases; the electron on the right which absorbs energy goes the opposite way. At each vertex, energy is not conserved, but if we consider the diagram as a whole, energy is conserved. If we observe the process from a distance, we don't see the virtual photon that is emitted and absorbed by the other electron because it violates energy conservation and can not be measured. So it is not a real process.

This diagram also contains other physics. For example, there is a time interval Δt, and an energy difference ΔE associated with the virtual photon. The product of ΔE and Δt is approximately $\hbar$, which is demanded by the uncertainty principle. It either indicates that there is a big energy violation within a very small time period or small energy violations over a long time period.

Another example is light absorption by a molecule as show in Figure 4.7.

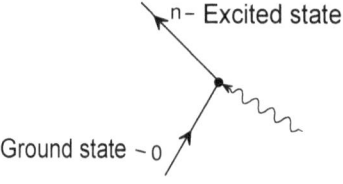

Figure 4.7: Molecular absorption of light

The molecule is initially in the ground state, and absorbs a photon causing a transition to an excited state. At some later time, the photon will be emitted and the system will return to the ground state. This process is not shown in the diagram.

We will no longer be concerned with the slopes of the molecule and photon lines in the diagrams because they will not affect the computational results. These diagrams will later be shown to be associated with specific terms in perturbation theory.

4.3.2 Elements of Feynman Diagram

The elements of the Feynman diagrams are:

- a line represents a molecule
- a wavy line represents a photon
- a vertex represents absorption (or emission) of a photon
- a diagonal line is a specific state
- a vertical line represents many possible states which do not necessarily conserve energy through a vertex

At the vertex, there is a virtual process, so energy does not have to be conserved. But energy is conserved within the whole process. These Feynman diagrams are just symbolic; they do not represent real particle trajectories. The vertical direction is time, and the horizontal direction is space but does not represent physical separations.

Next let's consider an example of linear susceptibility, draw all the possible diagrams as below:

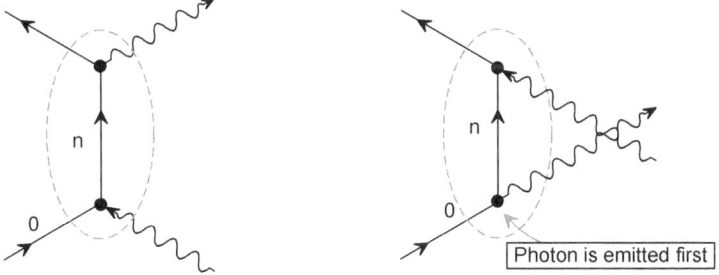

Figure 4.8: The two Feynman diagrams for the linear susceptibility

These two diagrams are equally likely, and they correspond to the two terms in the linear susceptibility $\chi_{ij}^{(2)}(-\omega_\sigma;\omega_\sigma)$. The fact that the photon is emitted before a photon is absorbed, as shown in the second figure, may seem strange because it is not a real process that is not really observed. For each case, energy is conserved for the whole process outside the dashed red regions.

Recall that

$$\chi_{ij}^{(2)}(-\omega_\sigma;\omega_\sigma) = -N\sum_p \left(\frac{\vec{\mu}_{0p}^j \vec{\mu}_{p0}^i}{\mathcal{E}_{p0}^* + \hbar\omega_\sigma} + \frac{\vec{\mu}_{0p}^i \vec{\mu}_{p0}^j}{\mathcal{E}_{p0} - \hbar\omega_\sigma} \right), \qquad (4.113)$$

where $\vec{\mu}_{0p}^k$ is the kth cartesian component of $\vec{\mu}_{0p}$, $\mathcal{E}_{p0} = \hbar\Omega_{p0}$, and where we have used the fact that the definition of $\chi_{ij}^{(2)}$ assigns i to the direction of the polarization $\vec{P}(\omega_\sigma)$, and j to the direction of the applied field $\vec{E}(\omega_\sigma)$. We will show in the next section how the two terms in Equation 4.113 are related to the two diagrams shown in Figure 4.8.

4.3.3 Rules of Feynman Diagram

Question:

How does one use Feynman diagrams to evaluate nonlinear susceptibilities?

We will use:

- Vertex rules

- Propagator rules

Vertex Rule: A vertex represents absorption or emission of a photon. In the dipole approximation the vertex represents a dipole transition between states of the system. If the initial state is $|0\rangle$ with one incident photon, then the dipole energy operator $\vec{\mu} \cdot \vec{E}$ couples the field to the final state $|n\rangle$ of the molecule with zero photons. Recall that $\vec{\mu}$ is an operator that acts on molecular state and $\vec{E}$ is an operator that acts on the state of the electromagnetic field. So, the expression for the vertex (Figure 4.9) gives a transition ampli-

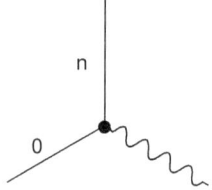

Figure 4.9: Example for vertex rule when photon is absorbed

tude $\langle n;0| \vec{\mu} \cdot \vec{E} |0;1\rangle$.

Recall that $\vec{E} \propto a + a^+$. So $\langle 0|\vec{E}|1\rangle \propto \langle 0|a|1\rangle = 1$. Then from the dipole moment of the molecule $\vec{\mu}$, we have $\langle n|ex_i|0\rangle$, which is simply expressed as ex_{n0}^i, where i is the field polarization direction.

Figure 4.10 shows a photon that is emitted first,

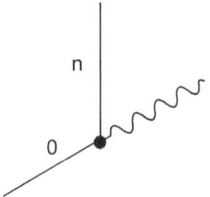

Figure 4.10: Example for vertex rule when photon is emitted

which yields the vertex contribution $\langle n;1|\vec{\mu}\cdot\vec{E}|0;0\rangle$.

Each vertex contributes to the Feynman diagram with terms of the form ex_{n0}^i.

Propagator Rule:

Time-dependent perturbation theory gives integrals of the form:

$$\int dt e^{-i\Omega_{n0}t} \langle f|\vec{\mu}\cdot\vec{E}|i\rangle. \tag{4.114}$$

Since the electric field $\vec{E}$ of light is proportional to $ae^{-i\omega t} + a^+ e^{i\omega t}$, the integrand includes a term of the form $e^{-i(\Omega_{n0}\pm\omega)t}$. Thus integration yields a factor proportional to $\frac{1}{\Omega_{n0}\pm\omega}$,

$$\int dt e^{-i\Omega_{n0}t}\langle f|\vec{\mu}\cdot\vec{E}|i\rangle = \int e^{-i(\Omega_{n0}\pm\omega)t}dt \propto \frac{1}{\Omega_{n0}\pm\omega} \tag{4.115}$$

Clearly, the product of the two vertices and propagator in Figure 4.8 yields the two terms in Equation 4.113.

Next, we consider the second-order case shown in Figure 4.11:

The Feynman diagram in Figure 4.11 gives the propagator

$$\frac{1}{\Omega_{m0} - \omega_1 - \omega_2}, \tag{4.116}$$

which is evaluated between the two crosses.

In summary, the **Propagator Rule** tells us about time evolution as given by the time integral.

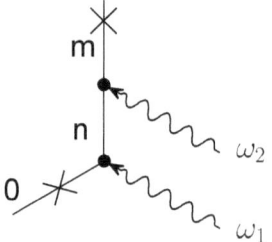

Figure 4.11: Example for propagator rule to second-order in the field.

4.3.4 Example: Using Feynman Diagrams to Evaluate Sum Frequency Generation

In the previous sections, we derived the mathematical expressions of the nonlinear susceptibilities using a time-consuming approach. In this section, we will see how we can write down all those mathematical expressions immediately just simply evaluating the Feynman Diagrams.

First, let's draw all possible diagrams for sum frequency generation (Figure 4.12). If we apply full permutation symmetry, we will get a total of 6 diagrams from this case (recall that 3! = 6).

Next, let's examine the first diagram which is drawn in more detail in Figure 4.13:

1. $|0\rangle$ to $|n\rangle$:

    - from the vertex rule, we have ex_{n0}^j
    - from the propagator rule, we have $\frac{1}{\hbar(\Omega_{n0}-\omega_1)}$

2. $|n\rangle$ to $|m\rangle$:

    - from the vertex rule, we have ex_{mn}^k
    - from the propagator rule, we have $\frac{1}{\hbar(\Omega_{m0}-\omega_1-\omega_2)}$, whre $\Omega_{m0} = \Omega_{mn} + \Omega_{n0}$. Note that in the denominator the energy difference is always from ground state to excited state.

3. $|m\rangle$ to $|0\rangle$:

    - from the vertex rule, we have ex_{0m}^i
    - from the propagator rule, since the transition is from the ground state back to the ground state, there is no energy denominator here.

Lecture Notes in Nonlinear Optics 109

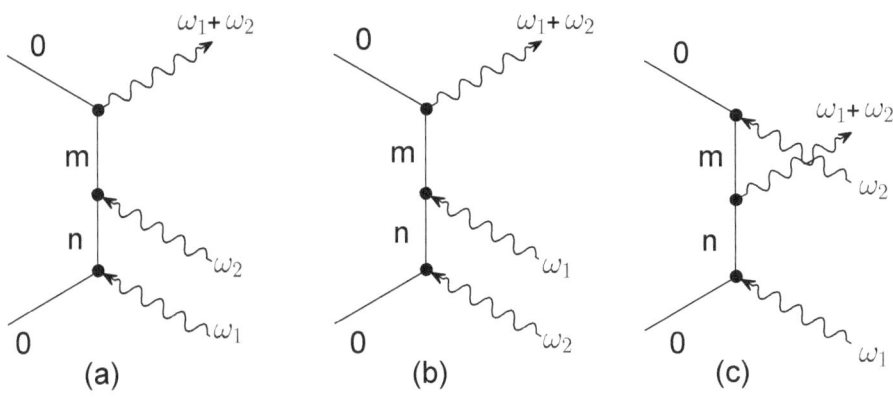

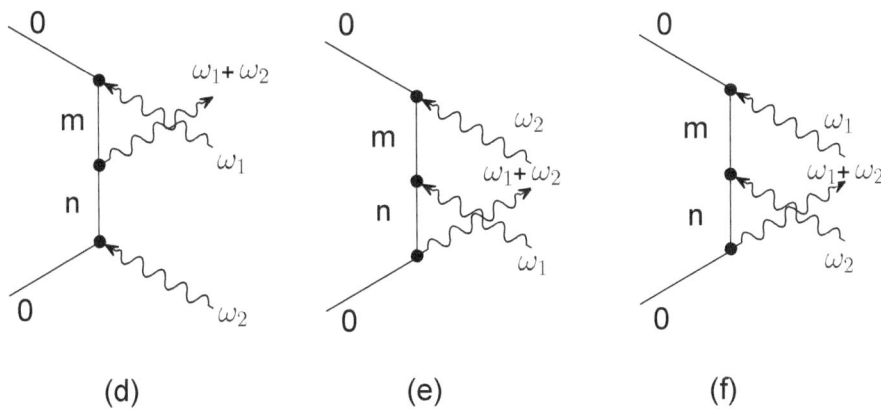

Figure 4.12: Feynman diagrams for sum frequency generation

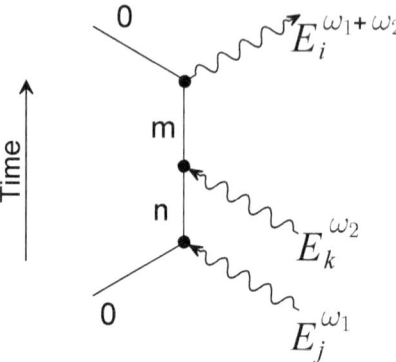

Figure 4.13: Sum frequency generation case a

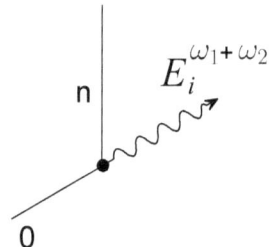

Figure 4.14: Outgoing photon in the propagator

We multiply all those terms together, which give us the contribution to β for that diagram

$$\beta = -\frac{e^3}{\hbar^2}\frac{x^i_{0m} x^k_{mn} x^j_{n0}}{(\Omega_{m0} - \omega_1 - \omega_2)(\Omega_{n0} - \omega_1)} \tag{4.117}$$

But Equation 4.117 is only for the first diagram. In order to get the total β, we need to add all the other five diagrams, which gives the result: β:

$$\beta = -\frac{e^3}{\hbar^2}\frac{x^i_{0m} x^k_{mn} x^j_{n0}}{(\Omega_{m0} - \omega_1 - \omega_2)(\Omega_{n0} - \omega_1)} + \textbf{the other five diagrams}. \tag{4.118}$$

It is important to mention that if the outgoing photon is in the propagator, Ω_{m0} is changed to Ω^*_{m0}. For example,

In Figure 4.14 the propagator rule gives $\frac{1}{\Omega^*_{n0} + \omega_1 + \omega_2}$. And all the other propagators above the one with the first complex conjugate are of the form Ω^*.

Lecture Notes in Nonlinear Optics 111

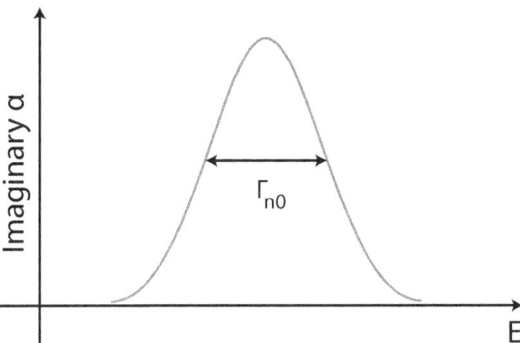

Figure 4.15: A plot of the imaginary part of α as a function of energy is used to demonstrate the decay width.

Problem 4.3-1: Using the Feynamn-like diagram below, calculate its contribution to $\chi^{(-)}_{----}(-;--\ldots)$, where you are to fill in the blanks. Note that the field at frequency $\omega_2 - \omega_1$ is the field that is being observed.

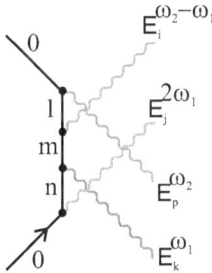

4.4 Broadening Mechanisms

Decay from an excited state to a lower state in the two-state system can be approximated by making it's Hamiltonian non-Hermitian with a small imaginary part of the energy,

$$E_{n0} \to E_{n0} - \frac{i\Gamma_{n0}}{2}, \qquad (4.119)$$

where Γ_{n0} is real and related to the inverse decay time constant through the uncertainty principle. Γ_{n0} can be represented as the energy width at half (or $1/e$) maximum of the imaginary part of α, as shown in Figure 4.15. The decay width can be directly measured with linear absorption spectroscopy. This is an example of homogeneous broadening mechanisms, where transition energy and lifetime of each molecule in the ensemble are related to each other with the uncertainty principle.

In nature, broadening is mainly inhomogeneous since molecules are found in different environments. As an example, consider a gas in which the velocities of the molecules obey the Boltzman distribution, i.e. they are moving with different velocities in different directions. Figure 4.16a shows the imaginary part of α versus energy for three particles, one that moves away from the observer, $v < 0$, one with zero velocity, $v = 0$, and one that is moving toward the observer, $v > 0$. The dashed curve represents the velocity distribution, which is used as a weighting factor in summing the individual peaks to determine the inhomogeneously-broadened spectrum.

As another example of inhomogeneous broadening, consider two molecules that collide and scatter so that the electric field, originating from Coulomb interactions, grows as a function of time when they get closer and decreases when they move away, leading to a shift in the eigenenergy due to the Stark effect. The position of the peak shifts continuously as the molecules approach each other. Figure 4.16b shows one such shift. The superposition of all the shifts, weighted according to the distribution of shifts from a scattering calculation, yields the inhomogeneous width.

The Doppler shift leads to the same kind of behavior. In the non-relativistic case, the shift in energy due to the Doppler effect is proportional to v/c. This implies that particles at rest with respect to the observer, $v = 0$, experience no Doppler shift. The effect of all the different processes should be taken into account when studying a quantum system. Figure 4.16c shows the superposition over all homogeneous-broadened molecules in the ensemble, which are weighted in amplitude to yield the dashed curve that represents the inhomogeneous spectrum.

In general, a quantum system is never isolated because it will interact with the environment. For example, when a molecule is in contact with a temperature bath, the energy has a Boltzman distribution. In the quantum picture, the molecules in vacuum are all in their ground states, but because of the temperature bath, other states may be populated as well. When the quantum system is not isolated and interacts with the environment, the system is characterized by both a quantum probability and an ensemble average weighted according to a statistical mechanical probability. In such cases, the density matrix is a convenient tool for understanding a system's properties.

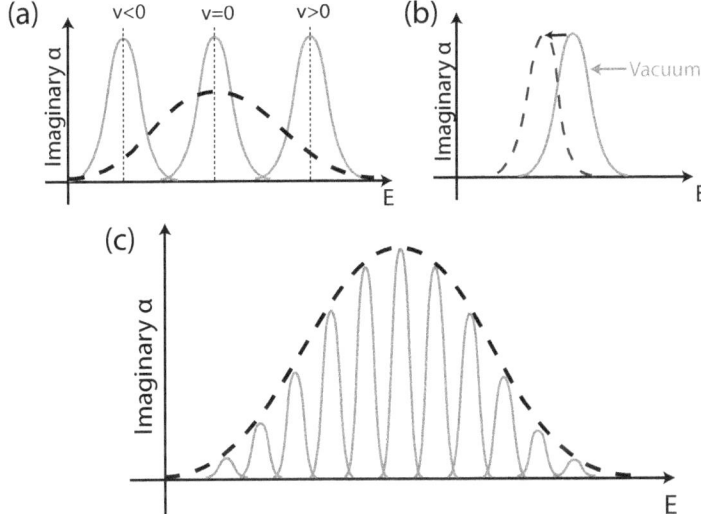

Figure 4.16: a) The absorption spectrum of three different molecules with different velocities (red), and the Boltzman distribution of peak positions (dashed curve). b) Shift in the spectrum of molecules that are interacting with each other. c) A schematic picture of the superposition of molecules in an ensemble weighted by the Boltzman factor, leading to a inhomogeneous broadening.

4.5 Introduction to Density Matrices

The density matrix, ρ, in Dirac notation is given by,

$$\rho = \sum_s p(s)|s(t)\rangle \langle s(t)|, \tag{4.120}$$

where summation is over states of the system, $|s(t)\rangle$, which are assumed to be normalized and $p(s)$ is the probability of finding the system to be in a state $|s\rangle$. The states $|s\rangle$ are not necessarily orthogonal and may or may not span the full Hilbert space. The consequence of this property will be described later. We stress that $p(s)$ is not necessarily a quantum probability. For example, when the system is in thermal equilibrium with a bath of temperature T, $p(s)$ would be proportional to $e^{-E_s/kT}$ where E_s is the energy of state s.

Density matrices describe a quantum system that interacts with the environment, where the interaction is taken into account via the probability term $p(s)$. The equilibrium condition determines which states are populated. If $|n\rangle$ is the eigenvector of the Hamiltonian,

$$H|n\rangle = E_n|n\rangle, \tag{4.121}$$

then any given state, $|s\rangle$, can be expanded as a superposition of energy eigenvectors, $|n\rangle$,

$$|s\rangle = \sum_n a_n|n\rangle, \tag{4.122}$$

where a_n is the *quantum* probability of finding the particle in state n and should not be confused with $p(s)$. a_n is the quantum amplitude and $p(s)$ is due to the interaction of the system with the outside world. We can calculate the quantum time evolution of state $|s\rangle$ via,

$$|s(t)\rangle = e^{-iHt/\hbar}|s\rangle = \sum_n a_n e^{-iE_n t/\hbar}|n\rangle. \tag{4.123}$$

Density matrices can be used to calculate the expectation value of any observable, A, by taking

$$\langle A\rangle = Tr[\rho A]. \tag{4.124}$$

Equation 4.124 takes into account both thermal and quantum effects. The meaning of Equation 4.124 becomes apparent by expressing it as

$$\begin{aligned}
Tr[\rho A] &= Tr\left[\sum_u p(u)|u\rangle\langle u|A\right] \\
&= \sum_u \sum_v \langle v|p(u)|u\rangle\langle u|A|v\rangle \\
&= \sum_{u,v} p(u)\langle u|A|v\rangle\langle v|u\rangle \\
&= \sum_u p(u)\langle A\rangle_{uu} \\
&= \langle A\rangle,
\end{aligned} \quad (4.125)$$

where we have used closure,

$$\sum_u |u\rangle\langle u| = 1. \quad (4.126)$$

Thus, the expected value of an observable is the sum over the quantum expectation values of that observable weighted by the probabilities of finding the system in those states.

The upper limit of the summation index in Equation 4.125 is the dimension of the Hilbert space, which for continuous variable is infinite. The dimensions of Hilbert space for a discrete systems such as spin 1/2 particles is 2: spin up, $|\uparrow\rangle$, and spin down, $|\downarrow\rangle$. The important point of the density matrix is that the sum in Equation 4.120 extends over the states of the system as it is initially prepared. When preparing the system, all the eigenstates may not be included or the system may include a state that is the superposition of many eigenstates, therefore the summation may span the Hilbert space, or may have more or less terms than the dimension of the Hilbert space. Examples will follow.

As a first example we consider a system of a spin half particle in a uniform magnetic field, $\mathbf{B} = B_0 \hat{z}$, in equilibrium with a bath at temperature T. The particle can be spin up or spin down. The energy of a particle with spin S_z is,

$$H = -\mu \cdot \mathbf{B} = -a S_z B_0, \quad (4.127)$$

where a is a constant and S_z is the spin component in z direction. The Hamiltonian in matrix form is

$$H = \begin{pmatrix} -a\hbar B_0/2 & 0 \\ 0 & a\hbar B_0/2 \end{pmatrix}, \quad (4.128)$$

and the probabilities of finding finding a particle with spin $\pm \hbar$ is,

$$p_\pm = \frac{e^{\pm a\hbar B_0/2kT}}{e^{a\hbar B_0/2kT} + e^{-a\hbar B_0/2kT}}. \quad (4.129)$$

Inserting Equation 4.129 in Equation 4.120 yields the density matrix,

$$\begin{aligned}\rho &= p_+|+\rangle\langle+| + p_-|-\rangle\langle-|\\ &= \begin{pmatrix} p_+ & 0 \\ 0 & p_- \end{pmatrix}.\end{aligned} \quad (4.130)$$

Now, using Equations 4.128 and 4.130 in Equation 4.124, the expectation value of the Hamiltonian can be calculated,

$$\langle H \rangle = Tr[H\rho] = -\frac{a\hbar B_0}{2}(p_+ - p_-) = -\frac{a\hbar B_0}{2}\tanh\frac{a\hbar B_0}{2kT}. \quad (4.131)$$

When the system is at zero temperature, the population is in its ground state, so Equation 4.131 yields, $\langle H \rangle = -a\hbar B_0/2$. Similarly, when the temperature becomes infinite, $\langle H \rangle = 0$. This makes sense because at infinite temperature both states are equally populated so that the average of the energies is zero.

As a second example, we describe a situation in which the number of states exceeds the dimension of the Hilbert space. Consider,

$$p(|+\rangle) = \frac{1}{4}, \quad p\left(\frac{|+\rangle + |-\rangle}{\sqrt{2}}\right) = \frac{1}{4}, \quad p(|-\rangle) = \frac{1}{2}. \quad (4.132)$$

So, there are 3 states in the population, which is larger than the dimensionality of the Hilbert space, which is 2. The density matrix is then given by,

$$\begin{aligned}\rho &= \frac{1}{4}|+\rangle\langle+| + \frac{1}{4}\left(\frac{|+\rangle + |-\rangle}{\sqrt{2}}\right)\left(\frac{\langle+| + \langle-|}{\sqrt{2}}\right) + \frac{1}{2}|-\rangle\langle-|\\ &= \begin{pmatrix} 3/8 & 1/8 \\ 1/8 & 5/8 \end{pmatrix}.\end{aligned} \quad (4.133)$$

The off-diagonal terms in the density matrix come from the quantum superposition of states. They are called coherence terms because they are non-zero only when the wave function is composed of a superposition of states. The trace of the density matrix given by Equation 4.133 is equal to unity. This is a general characteristic of density matrices.

For pure states, represented by a single vector in Hilbert space, the density matrix can be written as

$$\rho = |s(t)\rangle\langle s(t)|. \quad (4.134)$$

In this case,

$$\rho = \rho^\dagger \quad (4.135)$$

so a pure state density matrix must be Hermitian. Additionally, since the wavefunction is normalized, $\rho^2 = |s(t)\rangle \langle s(t)| \cdot |s(t)\rangle \langle s(t)| = |s(t)\rangle \langle s(t)| = \rho$. These two properties makes the density matrix for a pure state a projector.[1]

As an example, for a spin 1/2 particle, the most general real state vector is

$$|s\rangle = \cos(\theta)|+\rangle + \sin(\theta)|-\rangle, \qquad (4.136)$$

whence

$$\rho = \begin{pmatrix} \cos^2\theta & \sin\theta\cos\theta \\ \sin\theta\cos\theta & \sin^2\theta \end{pmatrix}. \qquad (4.137)$$

It is straightforward to show that this density matrix is a projector.

The time evolution of the density matrix operator can be studied in the Heisenberg representation, which for any operator A, yields,

$$\frac{dA}{dt} = \frac{\partial A}{\partial t} - \frac{i}{\hbar}[H, A]. \qquad (4.138)$$

For density matrix operator,

$$\frac{d\rho}{dt} = -\frac{i}{\hbar}[H, \rho], \qquad (4.139)$$

where we assume that the density matrix does not depend explicitly on time. The evolution of each component of ρ is given by

$$\frac{d\rho_{nm}}{dt} = -\frac{i}{\hbar}\langle n|H\rho - \rho H|m\rangle = -\frac{i}{\hbar}(E_n - E_m)\rho_{nm}, \qquad (4.140)$$

so, solving Equation 4.140, we see that the time evolution of the density matrix is given then by

$$\rho_{nm} = \rho_{nm}^{(0)} e^{-iE_{nm}t/\hbar}, \qquad (4.141)$$

where $\rho_{nm}^{(0)}$ is the density matrix at $t = 0$.

In matrix form, the expectation value of an operator A, $\langle A \rangle$, is calculated from the density matrix according to

$$\langle A \rangle = \sum_{n,m} \rho_{nm} A_{mn}. \qquad (4.142)$$

Using Equation 4.141, the time-dependent expectation value of the off-diagonal components, $n \neq m$, are found to oscillate at the Bohr frequency. For diagonal terms, $n = m$, the expectation value is time independent.

---

[1] see any textbook on quantum mechanics.

4.5.1 Phenomenological Model of Damping

In this section, we obtain an expression for the density matrix of a molecule that is not in equilibrium with its surroundings. When there is no interaction with the environment, all the particles are in their ground states, possessing the lowest energy. When particles interact with the environment, the system may get excited. So, the equilibrium state of the system will no longer be the ground state. The density matrix describes how the system is distributed, i.e. how many particles are in state $|a\rangle$, how many are in state $|b\rangle$ and so on. Consider a system interacting with light, where after being knocked out of the equilibrium, it decays back to equilibrium with a rate γ. The time evolution of the density matrix for this system can be modeled by

$$\frac{d\rho_{nm}}{dt} = -\frac{i}{\hbar}[H,\rho]_{nm} - \gamma_{nm}\left(\rho_{nm} - \rho_{nm}^{eq}\right), \qquad (4.143)$$

where ρ_{nm}^{eq} is the density matrix at equilibrium and the second term (corresponding to $\partial\rho/\partial t$), represents the relaxation rate when the system is out of equilibrium. Since γ represents the decay rate, γ_{nm} is a real quantity, which implies $\gamma_{nm} = \gamma_{mn}$. This means that a system that is underpopulated (depleted) will return back to the equilibrium at the same rate as if it is overpopulated. This implies a coherence between states n and m.

In general, a system can be in a superposition of multiple states. However, if we consider an ensemble, the particles must be in eigenstates and not in a superposition of states. For example, for a population of molecules in thermal equilibrium the energy is a parameter that determines the population, $\rho(E_n) \propto e^{-\beta E_n}$. The energy is in turn determined by the energy eigenvalues. Thermal effects will wash out coherence, i.e. for off-diagonal elements, $\rho_{nm}^{eq} = 0$. The population of each state is determined by the Boltzman factor, $e^{-E_n/kT}$, which are just the diagonal elements of the density matrix. Based on the discussion above, one can define γ_{nm} as a rate at which coherence between states n and m is lost.

The time evolution of ρ_{nn} is given by

$$\dot{\rho}_{nn} = -\frac{i}{\hbar}[H,\rho]_{nn} + \sum_{m>n}\Gamma_{nm}\rho_{mm} - \sum_{m<n}\Gamma_{mn}\rho_{nn}, \qquad (4.144)$$

where ρ_{nn} is the population of state n. When $m > n$, Γ_{mn} represents the decay rate from the higher states to the lower state n and when $n > m$, the decay rate represents transitions from n to the lower states. Thus, the sum over m includes all states from which n is populated (second term on the right-hand side of Equation 4.144), and to all the states to which state n can decay (the last term in Equation 4.144).

Lecture Notes in Nonlinear Optics 119

γ in Equation 4.143 is related to Γ in Equation 4.144. Consider a pure state, $|\psi(t)\rangle$, for which $\rho = |\psi(t)\rangle\langle\psi(t)|$. Its time dependent is given by,

$$|\psi(t)\rangle = e^{-iHt/\hbar}\sum_n a_n |n\rangle = \sum_n e^{-iE_n t/\hbar} a_n |n\rangle. \quad (4.145)$$

To include the decay of population, we can add a small imaginary part to the energy eigenstates given by Equation 4.119, so the density matrix becomes

$$\rho = \sum_{nm} a_m a_n^* e^{-i\omega_{mn}t} e^{-(\Gamma_m+\Gamma_n)t/\hbar} |m\rangle\langle n|, \quad (4.146)$$

whence

$$\rho_{nn} = |a_n|^2 e^{-\Gamma_n t}, \quad (4.147)$$

which indicates that ρ_{nn} decays with rate Γ_n. Taking the time derivative of Equation 4.147 yields,

$$\dot{\rho}_{nn} = -\Gamma_n \rho_{nn}. \quad (4.148)$$

Recall that Γ_n is the decay from state n. Comparing Equation 4.148 with Equation 4.144 leads us to the conclusion that

$$\Gamma_n = \sum_{m<n} \Gamma_{mn}. \quad (4.149)$$

On the other hand, associating the coefficient of $|m\rangle\langle n|$ in Equation 4.146 with ρ_{mn}, both $\sum_{m<n}\Gamma_{nm}\rho_{nn}$ and $\gamma_{nm}\left(\rho_{mm} - \rho_{nn}^{equil.}\right)$ constribute ρ_{nm}. Thus,

$$\gamma_{nm} = \frac{1}{2}(\Gamma_n + \Gamma_m) + \gamma_{nm}^{col}, \quad (4.150)$$

where the last term is due to collisions dephasing, where molecules collide without a gange in population; but, the coherence between the states n and m is lost.

To show the usefulness of the density matrix approach, consider a two-level atom where its dipole operator, μ, has the matrix form,

$$\mu = \begin{pmatrix} 0 & \mu_{ab} \\ \mu_{ba} & 0 \end{pmatrix}, \quad (4.151)$$

where a and b are the ground and excited states. Since the dipole moment operator is Hermitian, $\mu_{ab} = \mu_{ba}^*$. The expectation value of the dipole operator is given by

$$\langle\mu\rangle = Tr[\rho\mu] = Tr\left[\begin{pmatrix} \rho_{aa} & \rho_{ab} \\ \rho_{ba} & \rho_{bb} \end{pmatrix}\begin{pmatrix} 0 & \mu_{ab} \\ \mu_{ab}^* & 0 \end{pmatrix}\right] = \rho_{ab}\mu_{ab}^* + \rho_{ab}^*\mu_{ab}. \quad (4.152)$$

Equation 4.152 indicates that in the case of atoms, which have no dipole moment, coherence leads to a dipole moment: The dipole moment is induced in an atom that initially has no dipole moment when it is in contact with a thermal bath.

Furthermore, we can use perturbation theory to calculate the first order polarizability using the density matrix. The procedure is the same for the higher order nonlinearities. Given the Hamiltonian,

$$H = H_0 + \lambda V(t), \tag{4.153}$$

where $V(t)$ is the perturbation potential, we can use Equation 4.143 to study the evolution of the density matrix,

$$\dot{\rho}_{nm} = -\frac{i}{\hbar}(E_n - E_m)\rho_{nm} - \frac{i}{\hbar}[V(t),\rho]_{nm} - \gamma_{nm}\left(\rho_{nm} - \rho_{nm}^{eq}\right). \tag{4.154}$$

The density matrix can be expanded in a series of the order of the perturbation,

$$\rho = \rho^{(0)} + \lambda\rho^{(1)} + \cdots + \lambda^{(n)}\rho^{(n)} + \cdots. \tag{4.155}$$

Keeping the zeroth order terms of Equation 4.154 after substitution of Equation 4.155 leads to

$$\dot{\rho}_{nm}^{(0)} = -i\omega_{nm}\rho_{nm}^{(0)} - \gamma_{nm}\left(\rho_{nm}^{(0)} - \rho_{nm}^{eq}\right), \tag{4.156}$$

and to first order, $\lambda^{(1)}$,

$$\begin{aligned}\dot{\rho}_{nm}^{(1)} &= -i\omega_{nm}\rho_{nm}^{(1)} - \frac{i}{\hbar}\left[V(t),\rho^{(0)}\right]_{nm} - \gamma_{nm}\rho_{nm}^{(1)} \\ &= -(i\omega_{nm} + \gamma_{nm})\rho_{nm}^{(1)} - \frac{i}{\hbar}\left[\sum_p\left(V_{np}\rho_{pm}^{(0)} - \rho_{np}^{(0)}V_{pm}\right)\right], \end{aligned} \tag{4.157}$$

where we have used closure, which is given by Equation 4.126.

Because of thermal fluctuations, zeroth-order off-diagonal elements of the density matrix are zero in equilibrium, so only the diagonal terms in the summation in Equation 4.157 remain. Thus

$$\dot{\rho}_{nm}^{(1)} = -(i\omega_{nm} + \gamma_{nm})\rho_{nm}^{(1)} - \frac{i}{\hbar}V_{nm}\left(\rho_{mm}^{(0)} - \rho_{nn}^{(0)}\right), \tag{4.158}$$

where the last term is zero for $n = m$. When the perturbation term is zero, $V(t) = 0$, by definition $\rho_{nm} = \rho_{nm}^{eq}$, and it is zero when $n \neq m$. This implies that

$$\dot{\rho}_{nm}^{(0)} = 0 \quad for \quad n \neq m, \tag{4.159}$$

Lecture Notes in Nonlinear Optics 121

and
$$\dot{\rho}_{nn}^{(0)} = 0 \quad for \quad n = m, \quad (4.160)$$

which indicates that both diagonal and off-diagonal components are consistent with the zeroth order result.

Rearranging equation 4.157, we get

$$\dot{\rho}_{nm}^{(1)} + (i\omega_{nm} + \gamma_{nm})\rho_{nm}^{(1)} = -\frac{i}{\hbar}V_{nm}\left(\rho_{mm}^{(0)} - \rho_{nn}^{(0)}\right). \quad (4.161)$$

Equation 4.161 can be solved using Green's function as follows. Defining the operator, $\hat{O}_t$,

$$\hat{O}_t = \frac{\partial}{\partial t} + i\omega_{nm} + \gamma_{nm}, \quad (4.162)$$

Equation 4.157 can be written in the form,

$$\hat{O}_t \rho_{nm}^{(1)} = f(t) \quad (4.163)$$

where $f(t)$ is the right hand side of Equation 4.157. The general solution of Equation 4.163 is the homogeneous solution of $\rho_{nm,hom}^{(1)}$, for which $f(t) = 0$, plus the inhomogeneous solution, where $f(t) \neq 0$,

$$\rho_{nm}^{(1)} = \rho_{nm,hom}^{(1)} + \rho_{nm,inhom}^{(1)}. \quad (4.164)$$

We can easily find a solution for the homogeneous equation, yielding

$$\rho_{nm,hom}^{(1)} = Ae^{(-i\omega_{nm}+\gamma_{nm})t}, \quad (4.165)$$

where A is the integration constant. To solve the inhomogeneous equation, we define a Green's function, $G(t-t')$, so that,

$$\hat{O}_t G(t-t') = \delta(t-t'). \quad (4.166)$$

Once we have the Green's function, then we can find the inhomogeneous solution for the density matrix using the following relationship,

$$\rho_{nm,inhom}^{(1)}(t) = \int_{-\infty}^{\infty} G(t-t')f(t')dt'. \quad (4.167)$$

Clearly operating with $\hat{O}_t$ on Equation 4.167 with the use of Equation 4.170 yields $f(t)$,

$$\hat{O}_t \rho_{nm,inhom}^{(1)}(t) = \int_{-\infty}^{\infty} f(t')\hat{O}_t G(t-t')dt' = f(t), \quad (4.168)$$

where we have used the fact that

$$\int f(t')\delta(t'-t)dt' = f(t). \tag{4.169}$$

To solve a particular differential equation riquires a Green's function that can be constructed from a superposition of the homogeneous solutions. For a first order differential equation, the Green's function is simply a step function. We construct the Green's function by setting $A = 0$ for $t' < t$ and $A = 1$ in Equation 4.165 when $t' > t$. These two regions are connected through a step function where its derivative is a delta function. Thus

$$G(t'-t) = \begin{cases} 0 & t'-t < 0 \\ \exp\left[-(i\omega_{nm}+\gamma_{nm})t\right] & t'-t > 0. \end{cases} \tag{4.170}$$

Equivalently,

$$G(t'-t) = \Theta(t'-t)e^{-(i\omega_{nm}+\gamma_{nm})t}, \tag{4.171}$$

where $\Theta(t'-t)$ is a step function, with $\Theta(t'-t) = 0$ for $t'-t < 0$ and $\Theta(t'-t) = 1$ when $t'-t \geq 1$. Using Equation 4.171, the solution for $\rho_{nm}^{(1)}(t)$ in Equation 4.161 is obtained,

$$\rho_{nm}^{(1)}(t) = \frac{-i}{\hbar}\left[\int_t^\infty dt' e^{-(i\omega_{nm}+\gamma_{nm})t}V_{nm}\left(\rho_{mm}^{(0)}-\rho_{nn}^{(0)}\right)\right] + Ae^{-(i\omega_{nm}+\gamma_{nm})t}. \tag{4.172}$$

The expectation value of the dipole moment with

$$V = -\mu\cdot\mathbf{E} = -\mu\cdot\frac{\mathbf{E_0}}{2}e^{-i\omega_p t}+c.c., \tag{4.173}$$

and the first order density matrix,

$$\rho_{nm} = \rho_{nm}^{(0)} + \lambda\rho_{nm}^{(1)}. \tag{4.174}$$

eventually leads to a susceptibility,

$$\chi_{ij}^{(1)}(-\omega_p;\omega_p) = \frac{N}{\hbar}\sum_{n,m}\frac{\mu_{mn}^i\mu_{nm}^j}{\omega_{nm}-\omega_p-i\gamma_{nm}}\left(\rho_{mm}^{(0)}-\rho_{nn}^{(0)}\right). \tag{4.175}$$

Manipulating the index in Equation 4.175 gives an expression for $\chi^{(1)}$, which is very similar to what we had before,

$$\chi_{ij}^{(1)}(-\omega_p;\omega_p) = \frac{N}{\hbar}\sum_{n,m}\rho_{mm}^{(0)}\left[\frac{\mu_{mn}^i\mu_{nm}^j}{\omega_{nm}-\omega_p-i\gamma_{nm}}+\frac{\mu_{nm}^i\mu_{mn}^j}{\omega_{nm}+\omega_p+i\gamma_{nm}}\right]. \tag{4.176}$$

If we start from the ground state, $m = 0$, where we get $\rho_{00} = 1$, the expression for $\chi^{(1)}$ in Equation 4.176 is exactly as same as the results for vacuum where the particles are in the lowest energy state. However, in reality, the system can start from any other state. Therefore, $\rho_{mm}^{(0)}$ is a weight factor that describes the distribution of molecules in different states. Hence, the summation is over all possible initial states.

4.6 Symmetry

In this section we are going to use the symmetry of the Hamiltonian of a quantum system to investigate the properties of the nonlinear response. For a Hamiltonian given by

$$H = \frac{p^2}{2m} + V(r), \tag{4.177}$$

the energy eignestates, $|n\rangle$, are solutions of

$$H|n\rangle = E_n|n\rangle. \tag{4.178}$$

When the potential is symmetric,

$$V(\vec{r}) = V(-\vec{r}), \tag{4.179}$$

then the parity operator, π, commutes with the Hamiltonian $[H, \pi] = 0$, therefore

$$\pi|n\rangle = \pm|n\rangle. \tag{4.180}$$

As a result, the wave function can be characterized as either odd $|o_i\rangle$ or even $|e_i\rangle$ wave functions,

$$\begin{aligned}\pi|o_i\rangle &= -|o_i\rangle, \\ \pi|e_i\rangle &= +|e_i\rangle.\end{aligned} \tag{4.181}$$

In position space,

$$\langle \vec{x}|\pi|o_i\rangle = \langle -\vec{x}|o_i\rangle = -\langle \vec{x}|o_i\rangle, \tag{4.182}$$

which implies that the wave function is spatially asymmetric, $\psi_{oi}(-\vec{x}) = -\psi_{oi}(\vec{x})$.

We can apply these wave functions to determine the first order susceptibility,

$$\chi^{(1)}(\omega_p) = Ne^2 \sum_n |x_{0n}|^2 \left[\frac{1}{(E_{n0} - \hbar\omega_p)} + \frac{1}{(E_{n0}^* + \hbar\omega_p)} \right]. \tag{4.183}$$

Since x is an odd function under parity, $\langle e_i|x|e_j\rangle = \langle o_i|x|o_j\rangle = 0$, so no electric dipole moment exists for a centrosymmetric potential. Therefore, only transitions of the form $\langle e_i|x|o_i\rangle$ are allowed. According to Equation 4.183 there is a resonance in the susceptibility when the photon energy matches the transition energy, which results in a peak in the linear absorption spectrum at frequency $E_{i0}/\hbar$.

Usually, the ground state of a system with a symmetric potential is symmetric, which we will call $|e_0\rangle$. In this case only transitions to odd-parity excited states are allowed. These states are called one-photon states. On the other hand the even-parity states are called two-photon states because the transition moment from the ground state to a two photon state vanishes and so such states do not contribute to the first order susceptibility. Therefore, two-photon states are not observed in the linear absorption spectrum. However, as we will see later, transitions to two-photon states are observed in $\chi^{(3)}$ processes.

The contribution to the second order susceptibility from states 0 and 1, $\chi^{(2)}_{2L}$, is given by

$$\chi^{(2)}_{2L} \propto Ne^2 |x_{01}|^2 \Delta x_{10}, \qquad (4.184)$$

in which case both $|x_{01}|^2$ and Δx_{10} must be nonzero if state 1 is to contribute. For a centrosymmetric potential, when the first excited state is odd, $|x_{01}|^2 = \langle e_0|x|o_1\rangle \neq 0$. However, $\Delta x_{10} = x_{oo} - x_{ee} = \langle e|x|e\rangle - \langle o|x|o\rangle = 0$, and this results in $\chi^{(2)}_{2L} = 0$.

In the more general case that includes 3-level terms,

$$\chi^{(2)} \propto Ne^2 \sum_n |x_{0n}|^2 \Delta x_{n0} + Ne^2 \sum_{n,m} x_{0n} x_{nm} x_{m0}, \qquad (4.185)$$

where the first sum vanishes term-by-term according to the same argument as above. And $x_{0n} x_{nm} x_{mo}$ in the second term is always zero due to the fact that two of the three excited states are either both even or odd, which forces one of the matrix elements to vanish. Therefore, $\chi^{(2)}$ is always zero for a centrosymmetric potential.

In a two-level model, the third order susceptibility is of the form

$$\chi^{(3)}_{2L} \propto -|x_{01}|^2 (|x_{01}|^2 - |\Delta x_{10}|^2). \qquad (4.186)$$

Since the dipole moment is zero for a centrosymmetric potential, then $\chi^{(3)}_{2L} \propto |x_{01}|^4$. As we saw above, this term is nonzero. Thus, one strategy to make a good $\chi^{(3)}$ material is to use a centrosymmetric material that has a large $\chi^{(1)}$ value.

For potentials with no symmetry, we can always express the potential as a superposition of even- and odd-symmetry wave functions. In general we can write the nth state as

$$|n\rangle = \sin\alpha_n |n_e\rangle + \cos\alpha_n |n_o\rangle, \qquad (4.187)$$

Lecture Notes in Nonlinear Optics 125

where $\sin\alpha_n$ and $\cos\alpha_n$ are normalization coefficients. The susceptibilities for assymetric potentials can be expressed in terms of the the phase factors α_n. We leave the first order susceptibility derivation as an exercise. To write $\chi^{(2)}$ for a two-level system, we have to calculate $|x_{01}|^2$ and Δx_{10}. Using Equation 4.187, the wave functions are

$$|1\rangle = \sin\alpha_1 |1_e\rangle + \cos\alpha_1 |1_o\rangle,$$
$$|0\rangle = \sin\alpha_0 |0_e\rangle + \cos\alpha_0 |0_o\rangle. \quad (4.188)$$

Whence,

$$|x_{01}| = \langle 0|x|1\rangle = (\sin\alpha_0 \langle 0_e| + \cos\alpha_0 \langle 0_o|) x (\sin\alpha_1 |1_e\rangle + \cos\alpha_1 |1_o\rangle)$$
$$= \sin\alpha_0 \langle 0_e|x|1_o\rangle \cos\alpha_1 + \cos\alpha_0 \langle 0_o|x|1_e\rangle \sin\alpha_1, \quad (4.189)$$

and

$$\Delta x_{10} = x_{11} - x_{00} = \sin\alpha_1 \cos\alpha_1 \langle 1_e|x|1_o\rangle - \sin\alpha_0 \cos\alpha_0 \langle 0_e|x|0_o\rangle. \quad (4.190)$$

The only nonzero transitions are between states with opposite parities. The contribution to $\chi^{(2)}$ from these two states is determined by multiplying together Equation 4.189 and Equation 4.190. The result depends on the phases α_1 and α_0. If any of these angles are zero or π, then $\chi^{(2)}$ becomes zero. The optimum value of $\chi^{(2)}$ is between these two extremes.

4.7 Sum Rules

Now we demonstrate a simple approach to understand the magnitude of the nonlinear response using broad fundamental principles. Basically, to optimize a nonlinear response, one has to optimize infinite sums - a difficult if not impossible task. If a quantum system can be approximated by a two-level model, then a deeply-colored molecule, which necessarily has large $|x_{01}|$, will have a large linear susceptibility.

However, we can calculate an upper limit of the first order susceptibility using the sum rules. The sum rules, which are derived directly from the Schrodinger equation without approximation, state that

$$\sum_n |x_{n0}|^2 E_{n0} = N_{el} \frac{\hbar^2}{2m}, \quad (4.191)$$

where N_{el} is the number of electrons in the system and m is the mass of the electron. First we simplify Equation 4.183 by assuming that the light is in the off-resonant regime, so that we can approximate $\omega_p \approx 0$. Assuming $E_{n0} \sim E_{n0}^*$, yields,

$$\chi^{(1)} = Ne^2 \sum_n \frac{2|x_{0n}|^2}{E_{n0}}, \tag{4.192}$$

where N is the number density of molecules.

All terms in the sum in Equation 4.192 are positive and the energy denominator insures that each term, on average, gets smaller with increasing n. Multiplying each term in Equation 4.192 by E_{n0}/E_{n0} we get

$$\chi^{(1)} = Ne^2 \sum_n \frac{2|x_{0n}|^2 E_{n0}}{E_{n0}^2}. \tag{4.193}$$

Using the fact that $1/E_{n0} \leq 1/E_{10}$, we can replace E_{n0} by E_{10} in the denominator, leading to

$$\chi^{(1)} \leq Ne^2 \sum_n \frac{2|x_{0n}|^2 E_{n0}}{E_{10}^2}. \tag{4.194}$$

Factoring $1/E_{01}^2$ out from the sum yields

$$\chi^{(1)} \leq 2\frac{Ne^2}{E_{10}^2} \sum_n |x_{0n}|^2 E_{n0}, \tag{4.195}$$

and applying the sum rules gives

$$\chi^{(1)} \leq 2\frac{Ne^2}{E_{10}^2} \frac{N_{el} \hbar^2}{m}. \tag{4.196}$$

Equation 4.196 gives the fundamental limit for $\chi^{(1)}$. A similar derivation can be carried out for $\chi^{(2)}$ and $\chi^{(3)}$. Thus, for a given number of electrons and energy scale, there is a fundamental limit to all susceptibilities. These limits can be used to assess molecules and aid in the design of better materials.

4.8 Local Field Model

Figure 4.17a shows a material that is made of atoms or molecules. While the molecules are discrete, we are going to approximate the material surrounding

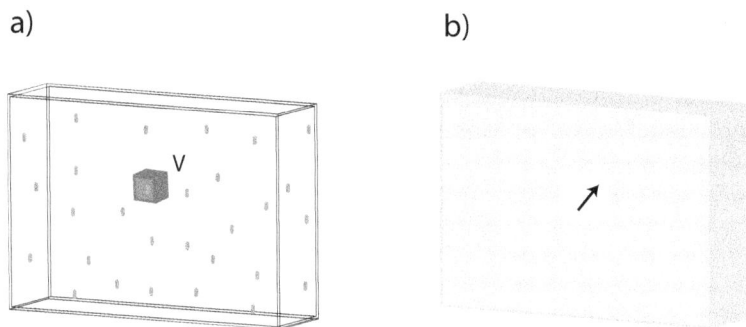

Figure 4.17: a) Discrete model: the material consists of atoms or molecules. b) Continuous model: The material is assumed a uniform dielectric and a molecule inside the material is treated as a discrete entity inside a cavity.

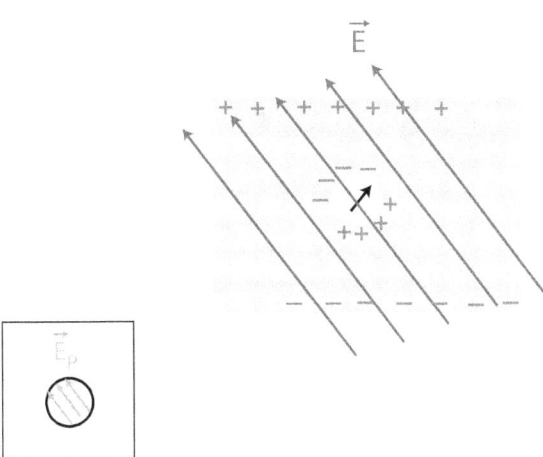

Figure 4.18: An electric field applied to a material induces charges on its surfaces. A molecule in the material, modeled as being inside a dielectric cavity (small arrow), is affected by both the applied field and the polarization field. The polarization field induced in the cavity is shown in the inset.

each molecule as being continuous, as shown in Figure 4.17b. Under this assumption we can describe the material by a dielectric constant ϵ, and assume that the molecule can be described as a discrete entity in a cavity.

Figure 4.18 illustrates the material in an electric field, $\vec{E}$. Charges are induced on the surface of the cavity, which creates a polarization field $\vec{E}_P$. In the discrete model, however, this polarization field that acts on a dipole is due to the neighboring dipoles.

The local field, $\vec{F}$, is the superposition of the applied field and polarization field,

$$\vec{F} = \vec{E} + \vec{E}_P. \tag{4.197}$$

In experiments, is is most convenient to measure the material response to the applied electric field rather than in terms of the local field. So, we need to develop a model that relates these two fields.[9] The Lorentz-Lorenz model is the simplest one, and assumes a uniform electric field, yielding

$$\vec{F} = \frac{\epsilon + 2}{3} \vec{E}. \tag{4.198}$$

For a plane wave of wave length λ, Equation 4.198 can be used when the cavity radius, a, is much smaller than the wavelength of the field ($a \ll \lambda$), as shown in Figure 4.19. The field is spatially uniform in the cavity but depends on time. The local field can then be expressed as a function of frequency

$$\vec{F}(\omega) = f(\omega)\vec{E}(\omega), \tag{4.199}$$

and

$$f(\omega) = \frac{\epsilon(\omega) + 2}{3}, \tag{4.200}$$

where $f(\omega)$ is the local field factor, and in general it is a tensor when the dielectric is anisotropic.

More complicated local field models consider the interaction between molecules, such as Onsager local field model, which includes the effects of the molecule on the dielectric. We are not going to discuss them here.

We can evaluate the local field factor using the relationship between the dielectric function and the refractive index, $\epsilon = n^2$. For visible light in glass, for example, $n \sim 1.5$, thus $f(\omega) \sim 1.42$. The local field in the glass is thus bigger than the applied field, $\vec{F}(\omega) > \vec{E}(\omega)$.

For an anisotropic material the generalized i^{th} component of the local field is,

$$F_i(\omega) = \sum_j f_{ij}(\omega) E_j(\omega), \tag{4.201}$$

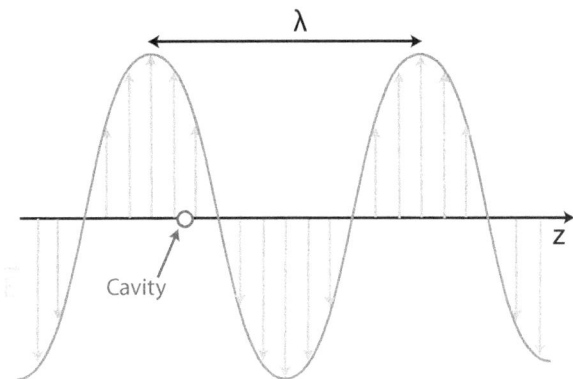

Figure 4.19: If the cavity radius is much smaller than the plane wave's wavelength, the electric field will be spatially uniform inside the cavity, which is necessary for the Lorentz-Lorenz local field model. The arrows are the electric fields.

in which $f(\omega)$ is a tensor. To apply this model to nonlinear optics, we can express the polarization in terms of the local fields

$$p_{i'}^{(n)}(\omega_\sigma) = \zeta_{i'jk\ldots}^{(n)}(-\omega_\sigma;\omega_1,\omega_2,\cdots)F_j(\omega_1)F_k(\omega_2)\cdots, \tag{4.202}$$

where $\zeta_{i'jk\ldots}^{(n)}$ is the nth order molecular susceptibility in vacuum. Equation 4.202 indicates that the induced dipole moment is due to the total field at the molecule. Substituting Equation 4.201 into Equation 4.202 yields

$$\begin{aligned}p_{i'}^{(n)}(\omega_\sigma) &= \zeta_{i'jk\ldots}^{(n)}(-\omega_\sigma;\omega_1,\omega_2,\cdots)f_{jj'}(\omega_1)f_{kk'}(\omega_2)\cdots \\ &\times E_{j'}(\omega_1)E_{k'}(\omega_2)\cdots,\end{aligned} \tag{4.203}$$

where summation convention holds. Multiplying both sides by $f_{ii'}(\omega_\sigma)$, and summing over i'

$$\begin{aligned}f_{ii'}(\omega_\sigma)p_{i'}^{(n)}(\omega_\sigma) &= \zeta_{i'jk\ldots}^{(n)}(-\omega_\sigma;\omega_1,\omega_2,\cdots)f_{ii'}(\omega_\sigma) \\ &\times f_{jj'}(\omega_1)f_{kk'}(\omega_2)\cdots E_{j'}(\omega_1)E_{k'}(\omega_2)\cdots.\end{aligned} \tag{4.204}$$

We define $p_{i'}^{(n)*}(\omega_\sigma)$ as the dressed dipole moment, and $\zeta_{i'jk\ldots}^{(n)*}(-\omega_\sigma;\omega_1,\omega_2,\cdots)$ as the dressed n^{th} order nonlinear molecular susceptibility, which are given by,

$$\begin{aligned}p_{i'}^{(n)*}(\omega_\sigma) &= f_{ii'}(\omega_\sigma)p_{i'}^{(n)}(\omega_\sigma), \\ \zeta_{i'jk\ldots}^{(n)*}(-\omega_\sigma;\omega_1,\omega_2,\cdots) &= \zeta_{i'jk\ldots}^{(n)}(-\omega_\sigma;\omega_1,\omega_2,\cdots) \\ &\times f_{ii'}(\omega_\sigma)f_{jj'}(\omega_1)f_{kk'}(\omega_2)\cdots.\end{aligned} \tag{4.205}$$

The dressed polarization can thus be expressed in terms of the dressed susceptibility and the applied fields,

$$p_{i'}^{(n)*}(\omega_\sigma) = \zeta_{i'jk\ldots}^{(n)*}(-\omega_\sigma;\omega_1,\omega_2,\cdots)E_{j'}(\omega_1)E_{k'}(\omega_2)\cdots. \quad (4.206)$$

These are the quantities that are measured by experiments.

The polarization is related to the dipole moment per unit volume, which can be calculated according to

$$P_i^{(n)}(\omega_\sigma) = \frac{1}{V}\sum_i p_i^{(n)*}(\omega_\sigma). \quad (4.207)$$

In Equation 4.207, the sum is over the dipoles in the given volume, V, shown in Figure 4.17a. On the other hand,

$$P_i^{(n)}(\omega_\sigma) = \chi_{ijk\ldots}^{(n)}(-\omega_\sigma;\omega_1,\omega_2,\cdots)E_j(\omega_1)E_k(\omega_2)\cdots. \quad (4.208)$$

Substituting Equation 4.206 into Equation 4.207 and subsequently into Equation 4.208 yields

$$\chi_{ijk\ldots}^{(n)}(-\omega_\sigma;\omega_1,\omega_2,\cdots) = \frac{N}{V}\zeta_{ijk\ldots}^{(n)*}(-\omega_\sigma;\omega_1,\omega_2,\cdots), \quad (4.209)$$

where N is the number of dipoles per volume.

We have assumed that all the dipoles are aligned in the same direction so that the sum over the dipole moments is simply given by the number of dipoles in the volume, N times the dipole moment of one molecule. If this assumption is not valid, then on the right hand side of Equation 4.209, we must replace $\zeta_{ijk\ldots}^{(n)*}$ with the orientational average of the molecular susceptibility $\langle \zeta_{ijk\ldots}^{(n)*} \rangle$ over the molecules in the unit volume, which is calculated using Equation 4.207. Thus, Equation 4.209 becomes

$$\chi_{ijk\ldots}^{(n)}(-\omega_\sigma;\omega_1,\omega_2,\cdots) = \frac{N}{V}\langle \zeta^{(n)*}(-\omega_\sigma;\omega_1,\omega_2,\cdots)\rangle_{ijk\ldots}, \quad (4.210)$$

It is worth mentioning that sometimes Ł is used to express the local field factor tensor in the form

$$Ł_{ii'jj'kk'\ldots}^{(n)}(-\omega_\sigma;\omega_1,\omega_2,\cdots) = f_{ii'}(\omega_\sigma)f_{jj'}(\omega_1)f_{kk'}(\omega_2)\cdots. \quad (4.211)$$

To illustrate the effect of the local field on the susceptibilities, we consider $\chi^{(3)}$, where $\chi^{(3)} \sim f^4$. The local field factor, for example, for visible light in glass is $f \simeq 1.42$, leading to a correction to γ by a factor of 4. Therefore, the local field factor results in an enhancement of the bulk nonlinear susceptibility relative to the sum over the microscopic units. Local fields can therefore be used to enhance the nonlinear optical response. For example, metal nanoparticles can act as strong electric field intensifiers that can theoretically lead to an enhancement of 8 orders of magnitude when using silver spheres at the surface plasmon resonance.

Chapter 5

Using the OKE to Determine Mechanisms

This chapter focuses on how the various mechanisms of the Optical Kerr Effect in a liquid affects the third-order susceptibility tensor; and, how measurements of the $\chi^{(3)}$ tensor can be used to study the underlying mechanisms. We begin with a general description of the mechanisms and end with a detailed discussion of how molecular reorientation and electronic mechanisms can be separated using polarized nonlinear refractive index measurements.

5.1 Intensity Dependent Refractive Index

Assume $|E(\omega)|^2 \gg |E(\omega')|^2$, where $E(\omega)$ and $E(\omega')$ are the pump and probe field, respectively. Since we are interested in measuring the pump beam, we focus on the polarization at frequency ω',

$$P^{(3)}_{\omega'} = \frac{3}{2}\chi^{(3)}\left(-\omega';\omega,-\omega,\omega'\right)\left(E_{\omega'}E_{\omega'}{}^* + E_{\omega}E_{\omega}{}^*\right)E_{\omega'}. \tag{5.1}$$

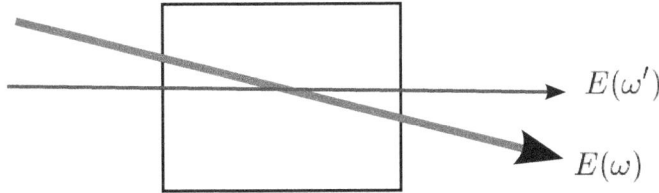

Figure 5.1: A pump beam of light influences the propagation of a probe beam.

We can use a color filter or beam block that only allows the beam at ω' to pass through to a detector. For weak pump fields, Equation 5.1 can be expressed to first order in $E_{\omega'}$,

$$P^{(3)}_{\omega'} = \frac{3}{2}\chi^{(3)}\left(-\omega';\omega,-\omega,\omega'\right)|E_\omega|^2 E_{\omega'}, \qquad (5.2)$$

which is the lowest order term that describes two-beam mixing. Note that the polarization in Equation 5.2 is proportional to the intensity of the strong beam.

To observe ω', we project out the ω' fourier component from Maxwell's wave equation. We obtain essentially the same result as when we did using a similar approach in Section 2.2. This leads to,

$$k^2_{\omega'} = \varepsilon(\omega')\frac{\omega'^2}{c^2} + 4\pi\frac{\omega'^2}{c^2}\frac{3}{2}\chi^{(3)}|E_\omega|^2, \qquad (5.3)$$

or

$$k^2_{\omega'} \equiv \varepsilon_{\text{eff}}\frac{\omega'^2}{c^2}, \qquad (5.4)$$

where $\varepsilon_{\text{eff}} = \varepsilon(\omega') + 6\pi\chi^{(3)}|E_\omega|^2$.

Then the effective refractive index is

$$n_{\text{eff}} = \sqrt{\varepsilon_{\text{eff}}} = n(\omega')\sqrt{1 + \frac{6\pi\chi^{(3)}}{n^2(\omega')}|E_\omega|^2}, \qquad (5.5)$$

where the refractive index is evaluated at ω'. For a small nonlinear phase shift $\chi^{(3)}|E_\omega| \ll 1$, the square root in Equation 5.5 can be expanded in a series,

$$n_{\text{eff}} \approx n(\omega') + \frac{3\pi\chi^{(3)}}{n(\omega')}|E_\omega|^2. \qquad (5.6)$$

This is similar to the one-beam result given by Equation 2.89. Recalling that

$$I_\omega = \frac{c}{8\pi}n(\omega)|E_\omega|^2, \qquad (5.7)$$

and substituting Equation 5.7 into Equation 5.6, the effective intensity-dependent refractive index becomes

$$n_{\text{eff}} \approx n(\omega') + \frac{24\pi^2\chi^{(3)}\left(-\omega';\omega,-\omega,\omega'\right)}{cn(\omega)n(\omega')}I_\omega. \qquad (5.8)$$

Lecture Notes in Nonlinear Optics 135

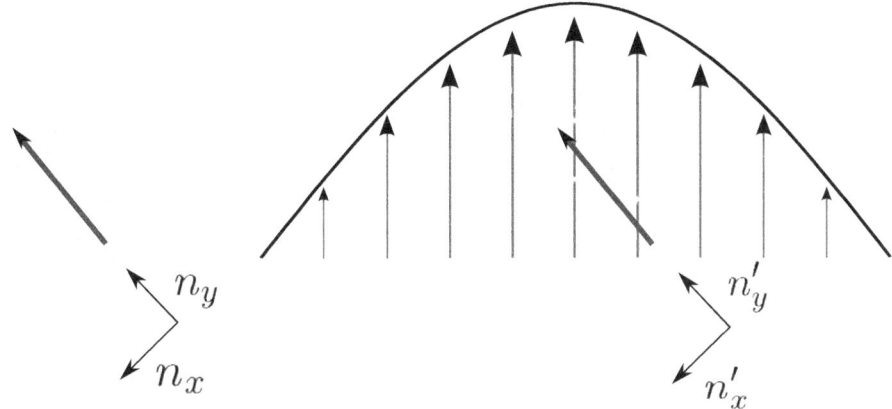

Figure 5.2: Electronic mechanism of $\chi^{(3)}$. A molecule with a nuclear framework represented by a blue arrow, and electrons represented by the surrounding green dots. The electron cloud is deformed by the electric field, as shown above. Note that the degree of deformation is greatly exaggerated.

The most general form of the intensity-dependent refractive index can be written as,

$$n = n_0 + n_2 I + n_4 I^2 + ..., \tag{5.9}$$

where $n_4 \propto \chi^{(5)}$, and $n_l \propto \chi^{(l+1)}$.

5.1.1 Mechanism of $\chi^{(3)}$

Electronic

Figure 5.2 shows a molecule and its electron cloud at an arbitrary initial orientation. As an electric field is applied, the shape of the molecule changes as the electrons move in response to the electric field. The refractive indices then change as the electron cloud changes. In practice the change in refractive index is measured by using the strong electric field of a very powerful laser pulse, and then probing the refractive index change with a weak laser.

Molecular Reorientation

A set of randomly oriented molecules is shown in Figure 5.3. The applied electric field is in the optical range, so oscillates too quickly for the molecules to keep up. The molecules will align axially along the polarization axis of the electric field as shown in Figure 5.3.

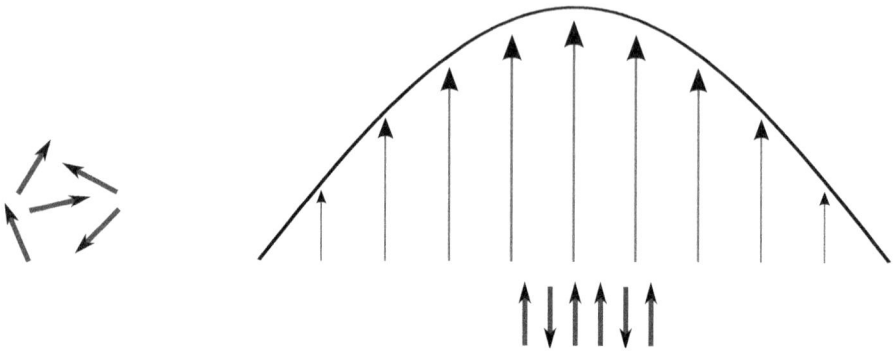

Figure 5.3: Reorientational mechanism of $\chi^{(3)}$. The nuclei of each molecule are shown as arrows.

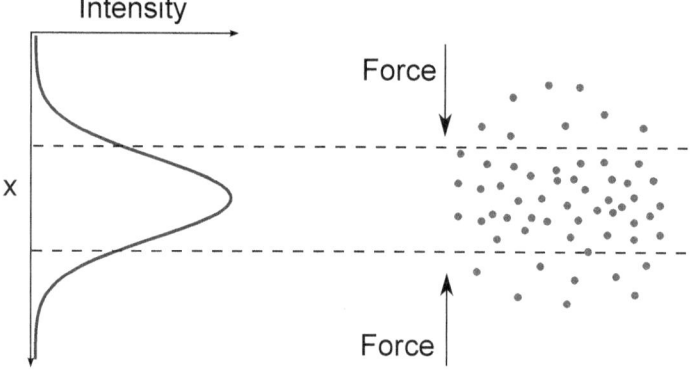

Figure 5.4: Electrostriction mechanism of $\chi^{(3)}$. The effective width of the light beam is shown by the dotted lines. The green dots represent molecules in solution.

Electrostriction(In liquid solution or gas phase)

A Gaussian beam, as shown in Figure 5.4, with full width half maximum shown by dashed lines, travels through a solution of molecules and creates an inward force on the molecules, which then move into the region of the beam. More molecules in a region results in a density change and thus a change in refractive index. In a solid material the force is on the material itself. For instance, rhythmic mechanical contraction and relaxation can be created by successively pulsing light into the material such that acoustical waves are produced.

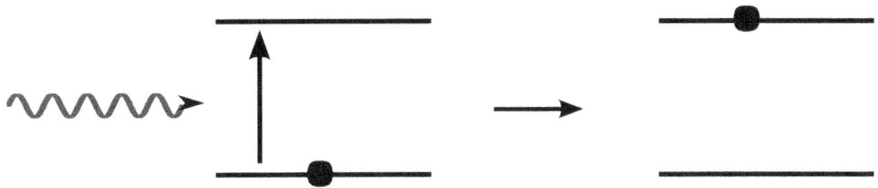

Figure 5.5: Saturated absorbtion mechanism of $\chi^{(3)}$.

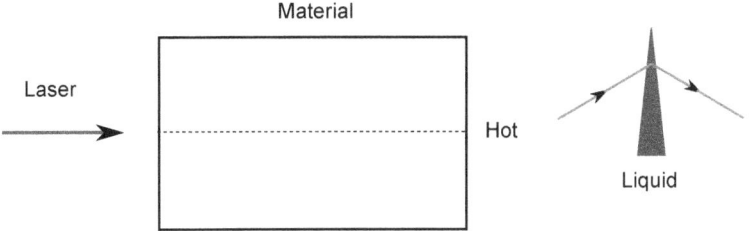

Figure 5.6: Thermal mechanism of $\chi^{(3)}$.

Saturated Absorbtion

In the electronic mechanism there is deformation of the electron cloud but there are no real excitations to excited states because the energy of the photon is much smaller than the transition energy in the molecule. However in saturated absorbtion, as shown in Figure 5.5, a molecule originally in an initial state is excited to a state of higher energy when the photon energy matches the transition energy, thus changing the refractive index.

Thermal Effect

When a laser impinges on a material as shown in Figure 5.6, it heats the material even in the off-resonance regime. In the region of the laser beam, this leads to a density change and a resulting refractive index change. Generally, these thermal effects have a negative refractive index.

An interesting example of a device that converts a continuous light beam to a pulsed source is a liquid placed in a thin prism cell. The prism bends the light, but through the thermal effect the liquid gets hot and the refractive index decreases. As the beam deflects due to heating, it moves away from the heated region, allowing the liquid to cool. Subsequently, the cooled liquid is back to its original state and the effect repeats. This leads to time-dependent periodic deflection of a continuous laser beam.

In Table 5.1, the strength of $\chi^{(3)}$ and the response time for each of the

Mechanism	$\chi^{(3)} \left(\frac{cm^3}{erg}\right)$	Response time (sec)
Electronic	10^{-14}	10^{-15}
Reorientation	10^{-12}	10^{-12}
Electrostriction	10^{-12}	10^{-9}
Saturated Absorbtion	10^{-8}	10^{-8}
Thermal	10^{-4}	10^{-3}

Table 5.1: $\chi^{(3)}$ and the response times of several mechanisms.

mechanisms are shown. There is a large change in the magnitude of $\chi^{(3)}$ from the electronic mechanism to the thermal mechanism; but, the ratio of the strength to the response time remains approximately constant. A device using the aforementioned mechanisms ideally has a large nonlinearity and also operates quickly. However, according to Table 5.1, if the device is fast $\chi^{(3)}$ is small so the device is less efficient and therefore requires a higher light intensity to operate. Thus a device operating under the different mechanisms, over any particular time scale, is essentially equivalently effective.

5.2 Tensor Nature of $\chi^{(3)}_{ijkl}$

The third-order susceptibility tensor, $\chi^{(3)}_{ijkl}$, is described by 81 separate elements. Without symmetry, many of these components can be independent and nonzero. However, when a material has a high degree of symmetry, the number of tensor components is significantly reduced. Liquids have centro-symmetric potentials and have vanishing $\chi^{(2)}$ so are thus ideal for studying third order susceptibility processes.

To understand the centro-symmetric behavior of liquids, it is convenient to define the inversion operator, $\hat{I}$, which changes the signs of all vector components. We also define $\hat{I}_i$, such that any vector becomes its inverse along only the ith component. In a liquid,

$$\hat{I}\chi^{(3)} = \chi^{(3)} \quad \text{and} \quad \hat{I}_i\chi^{(3)} = \chi^{(3)}, \qquad (5.10)$$

since if we rotate the liquid about any arbitrary axis it would look the same everywhere. The liquid is then said to be isotropic, i.e. the same in all directions; and homogeneous, i.e. the same in all places. For our purposes the isotropic behavior is particularly useful.

Consider
$$P_1^{(3)} = \chi_{1123}^{(3)} E_1 E_2 E_3. \qquad (5.11)$$

Applying $\hat{I}_2$ to Equation 5.11 yields,
$$\hat{I}_2\left[P_1^{(3)}\right] = \hat{I}_2\left[\chi_{1123}^{(3)} E_1 E_2 E_3\right], \qquad (5.12)$$
or
$$P_1^{(3)} = \chi_{1123}^{(3)} E_1(-E_2) E_3 = -\chi_{1123}^{(3)} E_1 E_2 E_3. \qquad (5.13)$$

Equations 5.11 and 5.13 reveal that $\chi_{1123}^{(3)} = -\chi_{1123}^{(3)}$, establishing that $\chi_{1123}^{(3)}$ must be zero. By applying similar operations, $\chi_{1122}^{(3)}$, $\chi_{1111}^{(3)}$, and other susceptibilities with even numbers of repeated components are nonzero. In total, we have 21 nonzero components of $\chi^{(3)}$ out of the original 81.

Rotating the system 90^0 about any particular axis does not change the physical properties of the material, thus
$$\chi_{1111} = \chi_{2222} = \chi_{3333}. \qquad (5.14)$$

Rotating any other component, say $\chi_{1212}^{(3)}$, 90^0 about any axis, should leave the material unchanged, so
$$\chi_{1212} = \chi_{1313}. \qquad (5.15)$$

Following this same logic, all $\chi_{iijj}^{(3)}$ are the same for $i \neq j$, all $\chi_{ijji}^{(3)}$ are the same for $i \neq j$, and all $\chi_{ijij}^{(3)}$ are the same for $i \neq j$. Thus,

$$\chi_{1111} = \chi_{2222} = \chi_{3333}, \qquad (5.16)$$
$$\chi_{1122} = \chi_{1133} = \chi_{2211} = \chi_{2233} = \chi_{3311} = \chi_{3322}, \qquad (5.17)$$
$$\chi_{1212} = \chi_{1313} = \chi_{2323} = \chi_{2121} = \chi_{3131} = \chi_{3232}, \qquad (5.18)$$
$$\chi_{1221} = \chi_{1331} = \chi_{2112} = \chi_{2332} = \chi_{3113} = \chi_{3223}. \qquad (5.19)$$

From the 21 nonzero terms, only four are independent components, three of which are off diagonal. We can write all the off-diagonal components in terms of 1's and 2's,
$$\chi_{ijkl}^{(3)} = \chi_{1122}^{(3)} \delta_{ij}\delta_{kl} + \chi_{1212}^{(3)} \delta_{ik}\delta_{jl} + \chi_{1221}^{(3)} \delta_{il}\delta_{jk}, \qquad (5.20)$$
since any other component can be determined from Equations 5.16-5.19.

Applying the Euler's Rotation Matrix to $\chi^{(3)}_{1111}$ by 45^0 should leave the liquid unchanged, because it is isotropic, and the results of the measurement should not change, thus

$$\chi^{(3)}_{1111} = \chi^{(3)}_{1122} + \chi^{(3)}_{1221} + \chi^{(3)}_{1212}. \tag{5.21}$$

Equation 5.20 was derived under the assumption $i \neq j$, but, by virtue of Equation 5.21, when $i = j$, Equation 5.20 still produces the same sum. The third order susceptibility tensor thus has only three independent tensor components in an isotropic substance.

Now we use permutation symmetry to reduce our independent components even further. In general the intensity dependent refractive index is of the form

$$\chi^{(3)}_{ijkl}(-\omega_2; \omega_1, \omega_2, -\omega_1), \tag{5.22}$$

and for the special case of one beam, we have,

$$\chi^{(3)}_{ijkl}(-\omega; \omega, \omega, -\omega). \tag{5.23}$$

If we exchange the j and k tensor components and the associated two degenerate frequencies under permutation symmetry, then,

$$\chi^{(3)}_{ijkl}(-\omega; \omega, \omega, -\omega) = \chi^{(3)}_{ikjl}(-\omega; \omega, \omega, -\omega), \tag{5.24}$$

and $\chi^{(3)}_{1212} = \chi^{(3)}_{1122}$. Therefore Equation 5.20 gives

$$\chi^{(3)}_{ijkl} = \chi^{(3)}_{1122}\left(\delta_{ij}\delta_{kl} + \delta_{ik}\delta_{jl}\right) + \chi^{(3)}_{1221}\delta_{il}\delta_{jk}. \tag{5.25}$$

Thus, the third order polarization with the appropriate degeneracy factor, is given by

$$P^{(3)} = \frac{3}{4}\chi^{(3)}_{ijkl}E_jE_kE_l^*. \tag{5.26}$$

Substituting the susceptibility from Equation 5.25 into Equation 5.26 yields

$$P^{(3)} = \frac{3}{4}\left[\chi^{(3)}_{1122}\left(E_j\left(\vec{E}\cdot\vec{E}^*\right)\delta_{ij} + \left(\vec{E}\cdot\vec{E}^*\right)E_k\delta_{ik}\right) + \chi^{(3)}_{1221}\left(\vec{E}\cdot\vec{E}\right)E_l^*\delta_{ie}\right]. \tag{5.27}$$

Defining $A = \frac{3}{2}\chi^{(3)}_{1122}$ and $B = \frac{3}{2}\chi^{(3)}_{1221}$, Equation 5.27 becomes

$$\vec{P}^{(3)} = A\left(\vec{E}\cdot\vec{E}^*\right)\vec{E} + \frac{1}{2}B\left(\vec{E}\cdot\vec{E}\right)\vec{E}^*. \tag{5.28}$$

It is useful to write P_i in terms of the effective nonlinear susceptibility,

$$P_i = \chi_{ij}^{eff} E_j, \tag{5.29}$$

where

$$\chi_{ij}^{eff} = \chi_{ij}^{(1)} + A'\left(\vec{E}\cdot\vec{E}^*\right)\delta_{ij} + \frac{1}{2}B'\left(E_i E_j^* + E_i^* E_j\right). \tag{5.30}$$

Here $A' = A - \frac{1}{2}B$ and $B' = B$. We can find the sum of the diagonal components using the trace

$$Tr\left(\chi_{ij}^{eff} - \chi_{ij}^{(1)}\right) = 3\left(A' + B'\right)E\cdot E^*. \tag{5.31}$$

In the next section we show how A' and B' are related to the mechanisms of $\chi^{(3)}$. In particular by measuring them we will be able to determine which mechanisms are playing an important role in any given process and on various time scales.

5.3 Molecular Reorientation

We begin by considering the process of molecular reorientation, which is slower than the electronic mechanism. We will assume the material to be composed of a collection of freely rotating anisotropic molecules. Common examples of such materials are liquids and gases made of anisotropic molecules with anisotropic polarization. Figure 5.7 is a diagram of a "cigar"-shaped molecule where the electric field is applied along the z-axis. Here, α is the polarizability tensor, $\vec{E}$ is the electric field, $\vec{P}$ is the polarization, and θ is the angle between the long axis of the molecule and the direction of the electric field. Note that the induced dipole moment, $\vec{p}$, is not necessarily along $\vec{E}$ or the principle axes of α.

An anisotropic molecule will tend to align with an applied electric field. Thus, the refractive index of a medium made of anisotropic molecules increases along the direction of the field, and the refractive index decreases perpendicular to the direction of the electric field. Molecular reorientation is due to an electric field induced torque, $\vec{\tau}$, which is given by

$$\vec{\tau} = \vec{p} \times \vec{E}. \tag{5.32}$$

It follows that the internal energy per molecule is,

$$U = -\vec{p}\cdot\vec{E}. \tag{5.33}$$

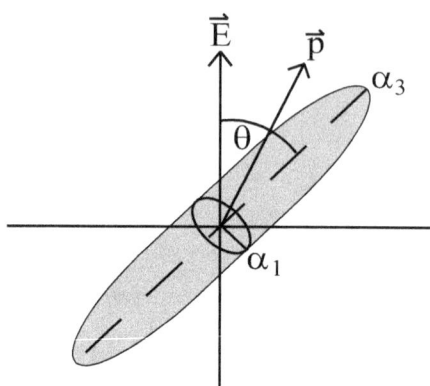

Figure 5.7: An anisotropic molecule in an electric field with anisotropic polarizability. Associated with the semi-major and semi-minor axes are the polarizabilities, α_3 and α_1, respectively.

Equation 5.33 shows that the energy is lowest when the molecule's long axis is aligned with the direction of the electric field. For a molecule that lacks a permanent dipole moment, there is no preferred polar orientation; that is, up and down orientation is equivalent. Figure 5.8 shows these lowest and highest energy configurations.

The polarization of an anisotropic molecule is given by

$$P_i = \alpha_{ij} E_j. \tag{5.34}$$

When the polarizability is anisotropic, the internal energy is found by integrating the right hand side of Equation 5.33. This gives,

$$U = -\int_0^{\vec{E}} \vec{p} \cdot d\vec{E}. \tag{5.35}$$

When a molecule is uniaxial as shown in Figure 5.7, the refractive index ellipsoid is described by two refractive indices. The refractive index component along the semi-minor axis is related to α_1 and refractive index component along the semi-major axis is related to α_3. Equation 5.35 for a uniaxial molecule can be evaluated with the help of Equation 5.34 and yields,

$$U = -\int_0^{\vec{E}} [\alpha_3 E_3' \, dE_3' + \alpha_1 E_1' \, dE_1']. \tag{5.36}$$

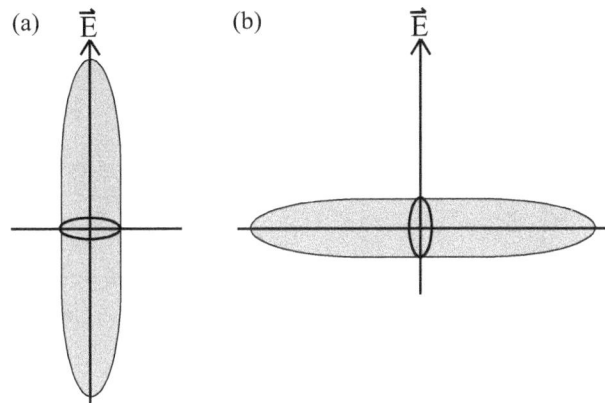

Figure 5.8: (a) The lowest orientational energy state; and, (b) the highest orientational energy state of an ellipsoidal anisotropic molecule in a uniform electric field.

E_3 and E_1 are the electric field components along the direction of the semi-major and semi-minor axis of the molecule, where $E_3 = E\cos\theta$ and $E_1 = E\sin\theta$. Upon integration, Equation 5.36 becomes,

$$U = -\frac{1}{2}E^2\left(\alpha_3 \cos^2\theta + \alpha_1 \sin^2\theta\right). \tag{5.37}$$

Figure 5.9 shows $U(\theta)$ given by Equation 5.37. We can rewrite this in a simplified form in terms of $\cos\theta$,

$$U = -\frac{1}{2}E^2\left[\alpha_1 + (\alpha_3 - \alpha_1)\cos^2\theta\right]. \tag{5.38}$$

The electric field in a plane wave of light is given by

$$E = E_0 \cos(\omega t), \tag{5.39}$$

where E_0 is the amplitude and ω is the angular frequency. Then the internal energy is proportional to $\cos^2(\omega t)$. By using a trigonometric identity, we can rewrite this as,

$$U = \frac{E_0}{4}\left[1 + \cos^2(2\omega t)\right]\left[\alpha_1 + (\alpha_3 - \alpha_1)\cos^2\theta\right]. \tag{5.40}$$

Since the reorientational response time is on the order of 10^{-12} seconds, any electric field of frequency greater than the reciprocal of the response time

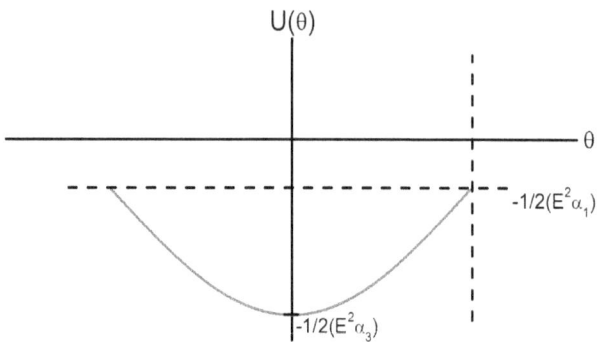

Figure 5.9: The energy per anisotropic molecule in an electric field, $\vec{E}$.

will not lead to significant reorientation. Since an optical cycle is on the order of 10^{-15} seconds and is therefore much shorter than reorientational time scales, we can take a time average of all quantities over an optical cycle. The time-averaged energy per molecule, U, is given by

$$\overline{U} = -\frac{1}{2}\overline{E^2}\left[\alpha_1 + (\alpha_3 - \alpha_1)\cos^2\theta\right]. \tag{5.41}$$

The linear refractive index is given by $n_{ii}^2 = 1 + 4\pi\chi_{ii}^{(1)}$. Note that there is no summation convention from double indices when the left hand side of the equation has those same indices. The refractive index is then given by the orientational ensemble average of the linear molecular susceptibility according to,

$$n_{ii}^2 = 1 + 4\pi N \langle \overleftrightarrow{\alpha} \rangle_{ii} \tag{5.42}$$

The linear molecular susceptibility in the lab frame, α_{zz}, can be found by rotating the molecule from the molecular frame. Thus,

$$\alpha_{zz} = \sum_{IJ} \alpha_{IJ} a_{Iz}(\Omega) a_{Jz}(\Omega), \tag{5.43}$$

where the a's are the rotation matrices and Ω represents the Euler angles θ, ϕ, and ψ.

We define $P(\Omega)$ as the probability density of finding a molecule at the angle Ω from Boltzmann statistics. Therefore,

$$P(\Omega) = \frac{e^{-\beta \overline{U}}}{\int d\Omega\, e^{-\beta \overline{U}}} \tag{5.44}$$

Lecture Notes in Nonlinear Optics 145

where $\beta = kT$ with k denoting the Boltzmann constant, and T denoting the temperature. Since no intermolecular forces are being considered in the energy, the orientational ensemble average of the linear molecular susceptibility is

$$\langle \overleftrightarrow{\alpha} \rangle_{ii} = \frac{1}{Z} \int d\Omega \alpha_{ii}(\Omega) e^{-\beta \overline{U}}. \tag{5.45}$$

where $d\Omega = d(\cos\theta) d\phi d\psi$. Z is the partition function defined as,

$$Z = \int d\Omega e^{-\beta \overline{U}}. \tag{5.46}$$

5.3.1 Zero Electric field

We will start by examining the susceptibility when no electric field is present. α_{xx} and α_{zz} are then given by Equation 5.43;

$$\alpha_{xx} = \alpha_1 + (\alpha_3 - \alpha_1)\sin^2\theta \cos^2\phi, \tag{5.47}$$

$$\alpha_{zz} = \alpha_1 + (\alpha_3 - \alpha_1)\cos^2\theta. \tag{5.48}$$

From Equation 5.41, when $\vec{E} = 0$, then $\overline{U} = 0$, then $P(\Omega) = 1$. Substituting Equations 5.47 and 5.48 into Equation 5.45, we get

$$\langle \overleftrightarrow{\alpha} \rangle_{xx} = \frac{\int_0^{2\pi} \int_{-1}^1 \int_0^{2\pi} d\psi\, d(\cos\theta)\, d\phi \left[\alpha_1 + (\alpha_3 - \alpha_1)\sin^2\theta \cos^2\phi\right]}{\int_0^{2\pi} \int_{-1}^1 \int_0^{2\pi} d\psi\, d(\cos\theta)\, d\phi}, \tag{5.49}$$

$$\langle \overleftrightarrow{\alpha} \rangle_{zz} = \frac{\int_0^{2\pi} \int_{-1}^1 \int_0^{2\pi} d\psi\, d(\cos\theta)\, d\phi \left[\alpha_1 + (\alpha_3 - \alpha_1)\cos^2\theta\right]}{\int_0^{2\pi} \int_{-1}^1 \int_0^{2\pi} d\psi\, d(\cos\theta)\, d\phi}. \tag{5.50}$$

After integration, the two non-zero linear molecular susceptibilities are,

$$\boxed{\langle \overleftrightarrow{\alpha} \rangle_{xx} = \frac{1}{3}\alpha_3 + \frac{2}{3}\alpha_1} \tag{5.51}$$

$$\boxed{\langle \overleftrightarrow{\alpha} \rangle_{zz} = \frac{1}{3}\alpha_3 + \frac{2}{3}\alpha_1} \tag{5.52}$$

Equations 5.51 and 5.52 are equivalent to one another as would be expected in an isotropic media consisting of anisotropic molecules which are randomly oriented. Therefore, the refractive index given by Equation 5.42, can be written as

$$n_0^2 \equiv n_{ii}^2 = 1 + \frac{4}{3}\pi N(\alpha_3 + 2\alpha_1). \tag{5.53}$$

5.3.2 Non-Zero Electric field

Next we consider the case when an electric field is present. When $\vec{E} \neq 0$, the time-averaged energy per molecule due to an external field is greater than zero. This means that the probability of finding a molecule at an angle Ω is no longer uniform. We define

$$J = \frac{1}{2}\overline{\beta E^2}(\alpha_3 - \alpha_1), \tag{5.54}$$

which is independent of the Euler angles and will thus remain after an integration is performed to calculate the ensemble average.

The orientational ensemble average of the polarizability will be used to determine the reorientational response. By, using Equations 5.48, 5.45, and 5.54, we get

$$\langle \overleftrightarrow{\alpha} \rangle_{zz} = \frac{\int_{-1}^{1} d(\cos\theta)[\alpha_1 + (\alpha_3 - \alpha_1)\cos^2\theta] e^{J\cos^2\theta}}{\int_{-1}^{1} d(\cos\theta) e^{J\cos^2\theta}}. \tag{5.55}$$

When the magnitude of the field is small enough so that the alignment energy is much smaller than thermal energies, we can expand the exponential term,

$$e^{J\cos^2\theta} \approx 1 + J\cos^2\theta + \cdots \tag{5.56}$$

and

$$\langle \overleftrightarrow{\alpha} \rangle_{zz} = \frac{1}{3}\alpha_3 + \frac{2}{3}\alpha_1 + \frac{2}{45}\overline{\beta E^2}(\alpha_3 - \alpha_1)^2. \tag{5.57}$$

Substituting Equation 5.57 into Equation 5.42, the refractive index in the z direction is

$$n_{zz}^2 = n_0^2 + \frac{8\pi N}{45}\overline{\beta E^2}(\alpha_3 - \alpha_1)^2. \tag{5.58}$$

We define $\Delta n = n_{zz} - n_0$. But $n_{zz}^2 - n_0^2 = (n_{zz} - n_0)(n_{zz} + n_0) \equiv (\Delta n)(2\bar{n})$. For a very small reorientationally-induced refractive index change, $\Delta n = \delta n_{zz}$, where $\bar{n} \approx n_0$, we can write

$$2n_0 \Delta n = \frac{8\pi N}{45}\overline{\beta E^2}(\alpha_3 - \alpha_1)^2. \tag{5.59}$$

Therefore,

$$\delta n_{zz} = \frac{4\pi N}{45 n_0}\overline{\beta E^2}(\alpha_3 - \alpha_1)^2. \tag{5.60}$$

Lecture Notes in Nonlinear Optics

Defining $\delta n_{zz} = n_2 \overline{E^2}$, and including the Lorentz local field correction $L(\omega)$, we can find the fourth rank tensor n_2^{zzzz} to be of the form,

$$n_2^{zzzz} = L^4(\omega) \frac{4\pi N}{45 n_0} \beta (\alpha_3 - \alpha_1)^2. \tag{5.61}$$

Then we can calculate n_2^{xxzz} using the relationship

$$n_2^{xxzz} = -\frac{1}{2} n_2^{zzzz}. \tag{5.62}$$

Thus, the refractive index increases along the field and decreases perpendicular to it, as expected.

5.3.3 General Case

In the most general case, there are three independent linear susceptibilities that can be represented by the diagonal elements of $\overleftrightarrow{\alpha}$,

$$\overleftrightarrow{\alpha} = \begin{pmatrix} a & 0 & 0 \\ 0 & b & 0 \\ 0 & 0 & c \end{pmatrix}. \tag{5.63}$$

The orientational ensemble average of $\overleftrightarrow{\alpha}_{ij}$ is the non-zero field result added to the zero field result. Thus, using the Kronecker delta function, δ_{ij},

$$\langle \alpha_{ij} \rangle = Q \delta_{ij} + \gamma_{ij} \left(\overline{E^2} \right), \tag{5.64}$$

where the first term on the righthand side of Equation 5.64 is from the generalized zero field solution and the second term is attributed to the third order susceptibility. Here, we state without proof that

$$Q = \frac{1}{3}(a + b + c), \tag{5.65}$$

where a, b, and c are the diagonal elements in Equation 5.63.

The second term on the righthand side of Equation 5.64 is calculated analogously to the derivative leading to Equation 5.61, yielding

$$\gamma_{ij} = C \left(3 \delta_{ik} \delta_{jl} - \delta_{ij} \delta_{kl} \right) E_k^{\text{loc}} E_l^{\text{loc}}. \tag{5.66}$$

Here, the superscript "loc" refers to the local electric field. The constant C is then of the form

$$C = \frac{\beta}{90} \left[(a-b)^2 + (b-c)^2 + (a-c)^2 \right]. \tag{5.67}$$

We can represent the electric field as,

$$E_k(t) = \frac{1}{2} E_k e^{-i\omega t} + \text{c.c.} \tag{5.68}$$

The polarization will be given by

$$P_i^{(3)} = N \sum_j \gamma_{ij} E_j. \tag{5.69}$$

Using Equations 5.66 - 5.68, the polarization can be found from Equation 5.69. In vector form, the polarization derived from Equation 5.69 is given by,

$$\vec{P}^{(3)} = A \left(\vec{E} \cdot \vec{E}^* \right) \vec{E} + \frac{1}{2} B \left(\vec{E} \cdot \vec{E} \right) \vec{E}^*, \tag{5.70}$$

where $A = NC/4$ and $B = 3NC/2$.

Recalling that

$$\frac{\delta n_{\text{linear}}}{\delta n_{\text{circular}}} = 1 + \frac{B}{2A}, \tag{5.71}$$

then the fraction $\frac{\delta n_{\text{linear}}}{\delta n_{\text{circular}}} = 4$. This is larger than the 3/2 of the electronic response, and therefore the two mechanisms are easily separable. Figure 5.10 shows a schematic diagram of the fraction $\frac{\delta n_{\text{linear}}}{\delta n_{\text{circular}}}$ as a function of laser pulse width.

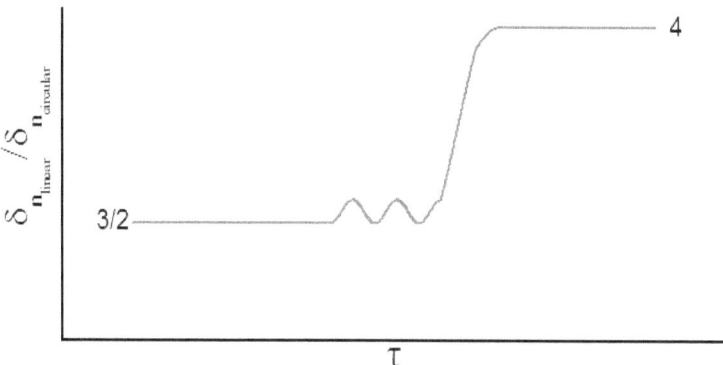

Figure 5.10: $\frac{\delta n_{\text{linear}}}{\delta n_{\text{circular}}}$ as a function of the laser pulse width τ.

This completes the full general description of molecular reorientation of ellipsoidal molecules.

5.4 Measurements of the Intensity-Dependent Refractive Index

In this section we will discuss the use of circularly or elliptically polarized light to measure the intensity-dependent refractive indices of a liquid to determine $\chi^{(3)}$. Using circular or elliptically polarized light, we can measure the tensor components without having to keep track of phase factors. Recall that we are considering the process governed by $\chi^{(3)}(-\omega;\omega,\omega,-\omega)$.

We begin by defining new unit vectors,

$$\hat{\sigma}_\pm = \frac{\hat{x} \pm i\hat{y}}{\sqrt{2}}, \tag{5.72}$$

that contain real and an imaginary components. Using the same notation as Boyd, $\hat{\sigma}_+$ describes left-hand circular polarization and $\hat{\sigma}_-$ describe right-hand circular polarization. Some properties of the unit vectors include:

$$\hat{\sigma}_\pm \cdot \hat{\sigma}_\pm = 0 \tag{5.73}$$
$$\hat{\sigma}_\pm^* = \hat{\sigma}_\mp \tag{5.74}$$
$$\hat{\sigma}_\pm \cdot \hat{\sigma}_\mp = 1. \tag{5.75}$$

The dot product of the unit vector with itself gives zero. This seems curious since unit vectors are orthonormal, but since these unit vectors contain a real and imaginary part, the dot product of the unit vector $\hat{\sigma}_\pm$ with itself gives zero. The complex conjugate of the unit vector $\hat{\sigma}_\pm$ gives the opposite unit vector, $\hat{\sigma}_\mp$. From these two properties, we can see that the dot product of a unit vector $\hat{\sigma}_\pm$ with its complex conjugate gives unity.

In electrodynamics, classical mechanics, and statistical mechanics, all quantities are real. In quantum mechanics, wavefunctions are complex, although measured values are always real. The complex quantities introduced here are used as a computational tool while the actual measurements will always yield real values.

The electric field can be written in terms of the unit vectors as a linear combination of right circular and left circular polarized light,

$$\vec{E} = E_+ \hat{\sigma}_+ + E_- \hat{\sigma}_-. \tag{5.76}$$

Some other properties of the electric field are

$$\vec{E} \cdot \vec{E} = 2E_+ E_- \tag{5.77}$$
$$\vec{E} \cdot \vec{E}^* = |E_+|^2 + |E_-|^2, \tag{5.78}$$

where $|E_+|^2$ and $|E_-|^2$ are the intensities of each of the two polarizations. The dot product of the electric field vector with itself thus gives a cross term. The dot product of the electric field vector times its complex conjugate gives the total intensity.

Last time we considered the nonlinear polarization and derived an equation for the nonlinear polarization in terms of right-circular and left-circular polarized light,

$$\vec{P}^{NL} = A(|E_+|^2 + |E_-|^2)\vec{E} + B(E_+ E_-)\vec{E}^*, \qquad (5.79)$$

where A and B are constants that can be calculated from the $\chi^{(3)}$ tensor. We can rewrite the nonlinear polarization in terms of the circular polarization unit vectors such that

$$\vec{P}^{NL} = P_+^{NL}\hat{\sigma}_+ + P_-^{NL}\hat{\sigma}_-, \qquad (5.80)$$

where P_+^{NL} contains the E_+ components of $\vec{E}$ and P_-^{NL} contains the E_- components of $\vec{E}$,

$$P_+^{NL} = A|E_+|^2 E_+ + A|E_-|^2 E_+ + B E_+ E_- E_-^* \qquad (5.81)$$

and

$$P_-^{NL} = A|E_+|^2 E_- + A|E_-|^2 E_- + B E_+ E_- E_+^*. \qquad (5.82)$$

Substituting Equations 5.81 and 5.82 into Equation 5.80 yields an Equation for $\vec{P}^{NL}$ in terms of $\hat{\sigma}_+$ and $\hat{\sigma}_-$,

$$\begin{aligned}\vec{P}^{NL} &= A|E_+|^2 E_+ \hat{\sigma}_+ + A|E_-|^2 E_+ \hat{\sigma}_+ + B E_+ E_- E_-^* \hat{\sigma}_+ \\ &\quad + A|E_+|^2 E_- \hat{\sigma}_- + A|E_-|^2 E_- \hat{\sigma}_- + B E_+ E_- E_+^* \hat{\sigma}_-.\end{aligned} \qquad (5.83)$$

Since $E_\pm E_\pm^* = |E_\pm|^2$, we can simplify Equation 5.83 by combining these terms,

$$\begin{aligned}\vec{P}^{NL} &= A|E_+|^2 E_+ \hat{\sigma}_+ + (A+B)|E_-|^2 E_+ \hat{\sigma}_+ \\ &\quad + A|E_-|^2 E_- \hat{\sigma}_- + (A+B)|E_+|^2 E_- \hat{\sigma}_-.\end{aligned} \qquad (5.84)$$

Recall that $P_\pm^{NL} = \chi_\pm^{NL} E_\pm$. In previous sections we have used χ^{eff}, but this is the same as χ^{NL}. We can solve for the nonlinear susceptibility term in terms of the squared electric field such that

$$\chi_\pm^{NL} = A|E_\pm|^2 + (A+B)|E_\mp|^2. \qquad (5.85)$$

This equation depicts the asymmetry of the refractive index for positive and negative circular polarizations.

From the nonlinear susceptibility, we can write the refractive index as a function of intensity,

$$n_\pm \approx n_0 + \frac{2\pi}{n_0}[A|E_\pm|^2 E_+ + (A+B)|E_\mp|^2 E_-], \quad (5.86)$$

where n_0 is the isotropic index of refraction. As previously shown, we have 2 independent tensors: χ_{1122} and χ_{1221}. Circularly polarized light probes both components and the fields induce a refractive index change. In other words, the field induces birefringence in the material. The left and right circularly polarized light probe different tensor components since there are different phase velocities for each polarization, and therefore both are required to measure all of the susceptibility tensor components. Linear polarized light is unaffected, because, if the sample is rotated, it remains unchanged by virtue of the fact that it is isotropic.

5.5 General Polarization

5.5.1 Plane Wave

The most general form of a plane wave written in terms of right and left circular polarization unit vectors is

$$\vec{E}(z) = A_+ \exp\left[\frac{in_+\omega z}{c}\right]\hat{\sigma}_+ + A_- \exp\left[\frac{in_-\omega z}{c}\right]\hat{\sigma}_-, \quad (5.87)$$

where A_+ and A_- are the amplitudes of the left and right circular polarized components. The components of the wave propagate at $\frac{c}{n_+}$ and $\frac{c}{n_-}$. This expression can be simplified by defining a phase factor and a k-vector,

$$\vec{E}(z) = \left(A_+ e^{i\theta}\hat{\sigma}_+ + A_- e^{-i\theta}\hat{\sigma}_-\right)e^{ik_m z}, \quad (5.88)$$

where

$$\theta = \frac{1}{2}\Delta n \frac{\omega}{c} z \quad (5.89)$$

$$k_m = \left(n_- + \frac{1}{2}\Delta n\right)\frac{\omega}{c} \quad (5.90)$$

$$\Delta n = n_+ - n_- = \frac{2\pi B}{n_0}\left(|E_-|^2 - |E_+|^2\right). \quad (5.91)$$

5.6 Special Cases

5.6.1 Linear Polarization

For the case of linear polarization, the left polarized electric field component is equal to the right polarized electric field component, $E_- = E_+$, and the refractive index difference is zero, that is $\Delta n = 0$. The amplitudes of the left and right polarization electric fields are equivilant, so $A_- = A_+$. The left and right circular components of the refractive indices are defined as,

$$n_\pm = \frac{2\pi}{n_0}\left[A|E_+|^2 + (A+B)|E_+|^2\right] + n_0. \tag{5.92}$$

Since the right and left polarized electric field components are equal, the intensity of light can be written as twice the intensity of the left polarization, $|E|^2 = 2|E_+|^2$. Substituting $\frac{1}{2}|E|^2$ for $|E_+|^2$ in Equation 5.92, we get

$$n_\pm = \frac{2\pi}{n_0}\left[\left(A + \frac{B}{2}\right)|E|^2\right] + n_0. \tag{5.93}$$

We can write the nonlinear term as the difference between the refractive indices of the left and right linear polarizations as

$$\delta n_{linear} = \frac{2\pi}{n_0}\left[\left(A + \frac{B}{2}\right)|E|^2\right] \tag{5.94}$$

$$= n_\pm - n_0. \tag{5.95}$$

5.6.2 Circular Polarization

For the case of left circular polarized light, the electric field for left circular polarized light is nonzero and the the electric field for right circular polarized light is zero, $E_+ \neq 0$ and $E_- = 0$. The general form for the electric field is

$$\vec{E}(z) = A_+ e^{\frac{in_+\omega z}{c}} \hat{\sigma}_+. \tag{5.96}$$

Using Equation 5.86, the nonlinear term for the change in refractive index for left and right polarization is

$$\delta n_{circular} = \frac{2\pi}{n_0}\left[A|E_\pm|^2\right] \tag{5.97}$$

$$= n_+ - n_0. \tag{5.98}$$

In order to construct the susceptibility tensor for a material, one needs two independent experiments. The first experiment uses linear polarized light to measure $n_\pm$ and we get $(A + \frac{B}{2})$. The second experiment uses circular polarization to measure $n_\pm$ from which we get A. Thus, we can use these two experiments to find the coefficients A and B.

Different mechanisms have different tensor properties. When measuring the susceptibility tensor for a material, one can take the ratio of different components to find the mechanisms responsible for the nonlinearities.

5.6.3 Elliptical Polarization

Consider the case of elliptically polarized light. The electric field vector can be written as Equation 5.88 but with $A_+ \neq A_-$, $A_+ \neq 0$, and $A_- \neq 0$. In the first term, we can write the unit vector $\hat{\sigma}_+$ in terms of x and y unit vectors as a function of phase angle,

$$\hat{\sigma}_+ e^{i\phi} = \frac{\hat{x} + i\hat{y}}{\sqrt{2}}(\cos\theta + i\sin\theta). \tag{5.99}$$

Expanding the $\hat{x}$ and $\hat{y}$ terms yields

$$\hat{\sigma}_+ e^{i\phi} = \frac{1}{\sqrt{2}}[\hat{x}\cos\theta - \hat{y}\sin\theta + i(\hat{x}\sin\theta + \hat{y}\cos\theta)]. \tag{5.100}$$

We can define the real terms and the imaginary terms as

$$\hat{x}' = \hat{x}\cos(\theta) - \hat{y}\sin(\theta) \tag{5.101}$$

and

$$\hat{y}' = \hat{x}\sin(\theta) + \hat{y}\cos(\theta). \tag{5.102}$$

Then we can relate the set of coordinates $\hat{x}, \hat{y}$ to the transformed set of coordinates $\hat{x}', \hat{y}'$ using a matrix rotation,

$$\begin{bmatrix} \hat{x}' \\ \hat{y}' \end{bmatrix} = \begin{bmatrix} \cos\theta & -\sin\theta \\ \sin\theta & \cos\theta \end{bmatrix} \begin{bmatrix} \hat{x} \\ \hat{y} \end{bmatrix}. \tag{5.103}$$

The axes $\hat{x}, \hat{y}$ represent the axes of the polarization ellipse of the beam. The same rotation can be applied to the unit vector $\hat{\sigma}_-$. The matrix transformation rotates the axes in a clockwise direction, as shown in Figure 5.11. In general, the net effect is that the axes of the ellipse rotate as the light propagates through the material. The light remains left and right polarized, but there is a phase shift. Physically, this represents rotation of the axes of the elliptical light as it propagates due to a phase shift between polarizations. Recall the refractive index difference from Equation 5.91 and the

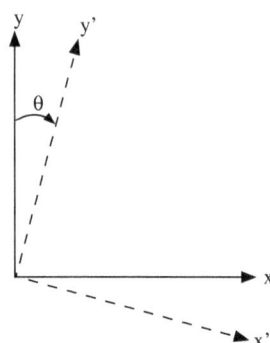

Figure 5.11: Rotation of axes $\hat{x}$ and $\hat{y}$ by an angle θ to give new axes $\hat{x}'$ and $\hat{y}'$.

change in the angle of axes from Equation 5.89 and notice that the rate of rotation depends on the difference in the refractive index. The degree of rotation depends on the amount of material the light propagates through. If we measure the change in angle of the axes, θ, we can find the difference in refractive index Δn, and obtain the constant, B.

5.7 Mechanisms

In order to calculate the tensor properties of the different mechanisms of $\chi^{(3)}$, we will consider how different mechanisms contribute to A and B. Different mechanisms will give rise to different ratios of A and B. So by measuring A and B and taking the relative ratios, we can determine the mechanisms causing changes in the refractive index.

5.7.1 Electronic Response

An electronic response comes from moving charges. Recall the classical anharmonic oscillator, for example, where the general three-dimensional potential energy can be written as

$$U(\vec{r}) = \frac{1}{2}m\omega_0^2|\vec{r}|^2 - \frac{1}{4}m^b|\vec{r}|^4. \tag{5.104}$$

In this case, the system is centrosymmetric, so $\chi^{(2)} = 0$. The solution for the general $\chi^{(3)}$ term in three dimensions is

$$\chi^{(3)}_{ijkl}(-\omega;\omega,\omega,-\omega) = \frac{Nbe^4\left[\delta_{ij}\delta_{kl} + \delta_{ik}\delta_{jl} + \delta_{il}\delta_{jk}\right]}{3m^3\mathscr{D}^3(\omega)\mathscr{D}(-\omega)}, \tag{5.105}$$

where $\mathcal{D}(\omega) = \omega_0^2 - \omega^2 - 2i\omega\Gamma$.

Recall that A and B in terms of the $\chi^{(3)}$ components are

$$A = 3\chi^{(3)}_{1122} + 3\chi^{(3)}_{1212} \qquad (5.106)$$

$$B = 6\chi^{(3)}_{1221}. \qquad (5.107)$$

For $\chi^{(3)}_{1122}$, only the first term in Equation 5.105 is nonzero. Similarly, for $\chi^{(3)}_{1212}$, only the second term is nonzero. Also, for $\chi^{(3)}_{1221}$, the last term is nonzero. But each of the terms in brackets in Equation 5.105 are unity, so this nonlinear oscillator model gives A = B. The ratio of the measurements δn_{linear} and $\delta n_{circular}$ yields

$$\frac{\delta n_{linear}}{\delta n_{circular}} = 1 + \frac{B}{2A} \qquad (5.108)$$

$$= \frac{3}{2}. \qquad (5.109)$$

For the electronic response mechanism, the ratio of δn_{linear} and $\delta n_{circular}$ is $\frac{3}{2}$. Notice that this ratio is independent of ω because the denominator is the same for all of the tensor components.

If a combination of $\chi^{(3)}$ mechanisms is acting, there will be a convolution of the ratios, so the mechanisms may not be simple to isolate. In this case we can use measurements of varying temporal pulse widths that probe the time scales characteristic of each mechanism.

5.7.2 Divergence Issue

Recall the general form of $\chi^{(3)}$ is written as

$$\chi^{(3)}_{kjih}(-\omega_\sigma; \omega_r, \omega_q, -\omega_p) = \frac{N}{\hbar^3} \mathcal{P}_F \sum_{lmn} \left[\frac{\mu^k_{0n} \mu^j_{nm} \mu^i_{ml} \mu^h_{l0}}{(\omega_{n0} - \omega_\sigma)(\omega_{m0} - \omega_q - \omega_p)(\omega_{l0} - \omega_p)} \right], \qquad (5.110)$$

where $\mathcal{P}_F$ is the full permutation symmetry operator over the frequencies ω_σ, ω_q, and ω_p and the sum is over all states, including the ground state. The outgoing frequency is a sum over the incoming frequencies,

$$\omega_\sigma = \omega_r + \omega_q + \omega_p. \qquad (5.111)$$

There is a divergence problem in the case where there is one beam such that $\omega_r \to \omega$, $\omega_q \to \omega$, $\omega_p \to -\omega$. In the ground state, $\omega_{00} = 0$. Adding two of the beams together yields zero, $-\omega_q - \omega_p = 0$, which diverges at all wavelengths. In order to solve the divergence problem, we would need to write out all 24 terms for $\chi^{(3)}$ which would yield two types of denominators of the general form,

$$\frac{1}{(X+Y)Y} + \frac{1}{(X+Y)X}. \tag{5.112}$$

When $X = -Y$, both terms diverge; however if we sum both terms, we get

$$\frac{1}{(X+Y)Y}\frac{X}{X} + \frac{1}{(X+Y)X}\frac{Y}{Y} = \frac{X+Y}{XY(X+Y)} \tag{5.113}$$

$$= \frac{1}{XY} \tag{5.114}$$

$$= \frac{-1}{X^2}, \tag{5.115}$$

so the sum of two diverging terms yields a well-behaved one.

5.7.3 One- and Two-Photon States

When we add all of the 24 terms and eliminate the divergence problem by combining divergent terms as described above, we get two different types of energy denominators,

$$\chi^{(3)}_{kjih}(-\omega_\sigma \; ; \; \omega_r, \omega_q, -\omega_p) =$$

$$\frac{N}{\hbar^3}\mathscr{P}_F \left[\sum_{lmn}' \frac{\mu^k_{0n}\mu^j_{nm}\mu^i_{ml}\mu^h_{l0}}{(\omega_{n0} - \omega_\sigma)(\omega_{m0} - \omega_q - \omega_p)(\omega_{l0} - \omega_p)} \right.$$

$$\left. - \sum_{ln}' \frac{\mu^k_{0n}\mu^j_{n0}\mu^i_{0l}\mu^h_{l0}}{(\omega_{n0} - \omega_\sigma)(\omega_{l0} - \omega_q)(\omega_{l0} - \omega_p)} \right]. \tag{5.116}$$

The $\sum'$ indicates a summation excluding the ground state. The first term is non-vanishing only when a two-photon state is included while the second term requires only one-photon states. If we assume that the incident photons are on or near resonance, then $\omega_{m0} - \omega_q - \omega_p \approx 0$ and $\omega_{l0} - \omega_p \approx 0$. Then, an incoming photon will match state n and will resonate.

The energy levels are illustrated in Figure 5.12 where the ground state, 0, is assumed to be of even parity; the state, n, is odd; and the state, m, is even. Additionally, n is a one-photon state such that transitions from 0 to n or n to m require one photon. The state m is a two-photon state such that one-photon cannot induce a transition from 0 to m.

From the quantum mechanical viewpoint, wave functions for centrosymmetric potentials have either even or odd parity. If the ground state is even,

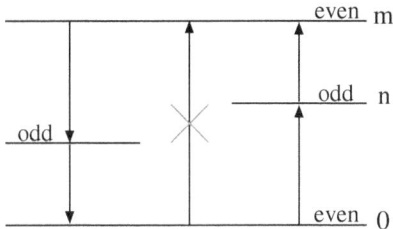

Figure 5.12: State n is a one-photon state where only one photon is required to induce a transition to this state. State m is a two-photon state where direct transitions from state 0 to state m are disallowed but transitions from states 0 to n and n to m are allowed.

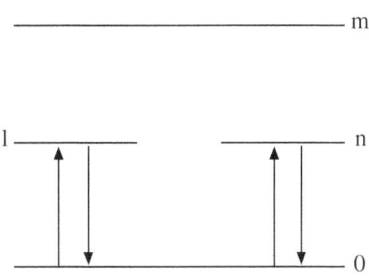

Figure 5.13: In the second term in Equation 5.116 only one-photon state transitions from the ground state, 0, to the first excited state, n or l, are allowed, but there are no two-photon states.

then the first transition state must be odd and the next transition state must be even. Therefore a transition from an even state to an even state is disallowed. In order to excite a transition to a state m, two photons are required where one photon induces a transition from the ground state to the odd-parity state and a second photon induces a transitions from the odd state to the even state. Both states n and l are one-photon states.

$\chi^{(3)}_{1111}$ will have both terms in Equation 5.116 where both one- and two-photon states are present. For $\chi^{(3)}_{1221}$, the first term in Equation 5.116 does not have to be zero but the second term must be zero. This can be illustrated in terms of wavefunctions inside a two-dimensional box as shown in Figure 5.14. In the most general case, the ground state will be an even state and then the first excited state is odd. Consider wavefunctions along x and y, $\psi_n(x)$ and $\psi_m(y)$. Then the centrosymmetric ground state can be written as

$$|\psi_1\rangle = |e_{x0}\rangle |e_{y0}\rangle, \qquad (5.117)$$

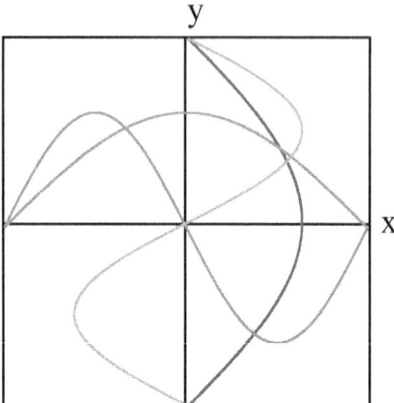

Figure 5.14: The ground state and first excited state wave functions in the x and y directions are depicted, where the ground state is even and the first excited state is odd. $|e_{x0}\rangle$ is red, $|o_{x1}\rangle$ is green, $|e_{y0}\rangle$ is blue, and $|o_{y1}\rangle$ is orange.

and the excited states can be written as

$$|\psi_2\rangle = |o_{xn}\rangle |o_{ym}\rangle, \tag{5.118}$$

where e and o refer to the even and odd parity state vectors, respectively. In the case where the first excited state is odd in the x-direction and even in the y-direction, the wave function,

$$|\psi_2\rangle = |o_{xn}\rangle |e_{y0}\rangle. \tag{5.119}$$

has a nonzero x-transition moment between the ground state and the first excited state,

$$\langle \psi_2 |x| \psi_1 \rangle \neq 0, \tag{5.120}$$

which we can see as follows.

We can use the wave functions defined in Equations 5.117 and 5.119 to calculate the transition moment,

$$\langle o_{xn}|\langle e_{y0}|x|e_{x0}\rangle|e_{y0}\rangle. \tag{5.121}$$

This can be simplified since the y states do not act on x, so we can move them outside of the bras and kets and then using $\langle e_{y0}|e_{y0}\rangle = 1$, because the states are orthonormal, we get,

$$\langle o_{xn}|x|e_{x0}\rangle \neq 0. \tag{5.122}$$

Lecture Notes in Nonlinear Optics 159

This represents a transition from an even state to an odd state.

If we use these state vectors that give nonzero transitions in the x-direction to calculate the transition moment of y, we get

$$\langle \psi_2 | y | \psi_1 \rangle = 0, \tag{5.123}$$

because

$$\langle \psi_2 | y | \psi_1 \rangle = \langle 0_{xn} | \langle e_{y0} | y | e_{x0} \rangle | e_{y0} \rangle. \tag{5.124}$$

Since the x operators do not act on y, we can move them outside the other bras and kets and are left with

$$\langle e_{y0} | y | e_{y0} \rangle = 0. \tag{5.125}$$

This expectation value vanishes because the bra, ket, and y operator are all odd in the y-direction.

Equation 5.125 shows that $\chi^{(3)}_{1221}$ cannot have two-photon character, but only have one-photon character. Thus, $\chi^{(3)}_{1221}$ has only two-photon character while $\chi^{(3)}_{1122}$ has both one- and two-photon character. The ratio of the $\chi^{(3)}$ components, $\frac{\chi^{(3)}_{1221}}{\chi^{(3)}_{1122}}$ is frequency-dependent so we cannot uniquely determine A and B, and therefore, we cannot learn about the mechanisms using any combination of polarized measurements.

In the off-resonance case, $\omega \to 0$ so that all frequencies vansish. Then, the two-photon terms are greater than or equal to zero and the one-photon terms are less than or equal to zero. Using Klienman symmetry, where we permute only the two interior wavelengths, then A = B. In this case, the ratio of the change in the linear refractive index to the change in the circular refractive index is equal to $\frac{3}{2}$,

$$\frac{\delta n_{linear}}{\delta n_{circular}} = 1 + \frac{B}{2A} \tag{5.126}$$

$$= \frac{3}{2}. \tag{5.127}$$

Thus, polarization-dependent measurements that yield *A* and *B* can be used to determine if the electronic or reorientational mechanisms are responsible.

Chapter 6

Applications

6.1 Optical Phase Conjugation

Let us consider a plane wave incident on a mirror. The incident plane wave field is given by

$$\vec{E}_{in} = \hat{\epsilon}\frac{A(\vec{r})}{2}e^{i(kz-\omega t)} + c.c. \tag{6.1}$$

The polarization $\hat{\epsilon}$ can be either circular, linear, or elliptical. From our boundary conditions the field at the surface must be zero

$$(\vec{E}_r + \vec{E}_{in})|_{z=0} = 0, \tag{6.2}$$

where $\vec{E}_r$ and $\vec{E}_{in}$ are the reflected and incoming fields respectively. Because the reflected wave moves opposite to the incident wave we find

$$\vec{E}_r = -\hat{\epsilon}\frac{A(\vec{r})}{2}e^{i(-kz-\omega t)} + c.c., \tag{6.3}$$

which is similar to the incoming wave where the polarization has been reversed and the wave vector is in the negative direction. We can incorporate the spatial portion of the exponential into $A(\vec{r})$ and we find

$$\vec{E} = \hat{\epsilon}\frac{A(\vec{r})}{2}e^{-i\omega t} + c.c. \tag{6.4}$$

The phase conjugate wave is defined as

$$\vec{E}_c(\vec{r},t) = \hat{\epsilon}^*\frac{A^*(\vec{r})}{2}e^{-i\omega t} + c.c, \tag{6.5}$$

$$= \hat{\epsilon}\frac{A(\vec{r})}{2}e^{i\omega t} + c.c, \tag{6.6}$$

$$= \vec{E}(\vec{r},-t). \tag{6.7}$$

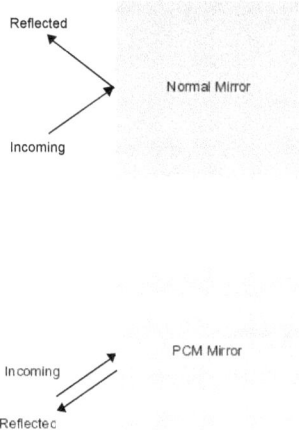

Figure 6.1: In a phase conjugate mirror the beam retraces the original path in reverse.

Where equation 6.6 is obtained by exchaning the two terms in equation 6.5. Phase conjugation is thus equivalent to time reversal, where $t \to -t$. The difference between reflection from a regular mirror and a phase conjugate mirror is shown in Figure 6.1 and Figure 6.2

6.2 Phase Conjugate Mirror

A phase conjugate mirror can be made using degenerate four wave mixing, a process that is mediated by $\chi^{(3)}$. We begin with two pump beams (E_1, E_2), a probe beam E_3 and the reflected beam E_4 (see figure 6.3). The source of phase conjugation can be understood by recognizing that the probe beam and one of the pump beams interfere and form a "grating" from which the other pump scatters. The scattered pump beam then travels in the direction of the phase conjugate beam, as shown in figure 6.3

For our derivation we assume the pump beams are much more intense than the probe and reflected beams, therefore we can assume that pump beam depletion is negligible. We will also assume that the fields are scalars, or,

$$\vec{E}_i(\vec{r}, t) \Rightarrow \frac{A(\vec{r})}{2} e^{i(\vec{k}_i \cdot \vec{r} - \omega t)} + c.c.. \tag{6.8}$$

Note that the polarization of the fields determine which components of the $\chi^{(3)}$ tensor are responsible, so long as $\chi^{(3)} \neq 0$ for the components of interest,

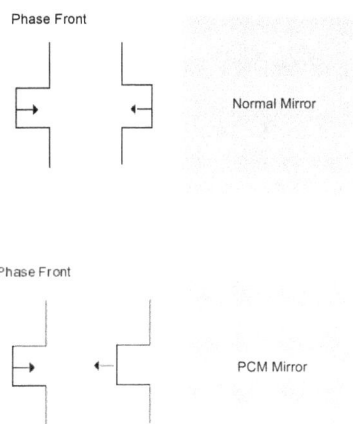

Figure 6.2: When a wavefront is incident on a mirror, the reflected wave's phase front is reversed. In a phase conjugate mirror the phase front is the same as that of the incident wave.

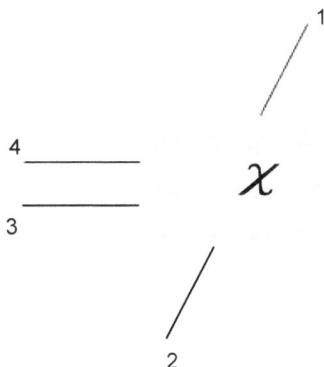

Figure 6.3: In a phase conjugate mirror, two counter propogating pump beams(1,2) interact with a probe beam(3) to create a phase conjugate(4).

the scalar approximation does not limit the generalility of the calculation. From the wave equation we know

$$E_4 \propto (P^{NL(3)}) = \frac{3}{2}\chi^{(3)} E_1 E_2 E_3^*. \tag{6.9}$$

Substituting the fields from equation 6.8 into equation 6.9,

$$E_4 \propto \frac{3}{2}\chi^{(3)} A_1 A_2 A_3^* e^{i(\vec{k}_1+\vec{k}_2-\vec{k}_3)\cdot\vec{r}}. \tag{6.10}$$

Where the two pump beams are counter propogating $\vec{k}_1 = -\vec{k}_2$, so equation 6.10 becomes

$$E_4 \propto \frac{3}{2}\chi^{(3)} A_1 A_2 A_3^* e^{-i\vec{k}_3\cdot\vec{r}}, \tag{6.11}$$

which shows that $E_4 \propto E_3^*$ making it the phase conjugate wave. Note that if A_1 and A_2 are sufficiently large, E_4 can be an amplified version of E_3.

To solve the problem in detail we consider the polarizations to first order in E_3 and E_4 as they are much smaller than E_1 and E_2. The third order polarizations are

$$P_1 = \frac{3}{2}\chi^{(3)}\left[E_1^2 E_1^* + 2E_1 E_2 E_2^*\right], \tag{6.12}$$

$$P_2 = \frac{3}{2}\chi^{(3)}\left[E_2^2 E_2^* + 2E_2 E_1 E_1^*\right], \tag{6.13}$$

$$P_3 = \frac{3}{2}\chi^{(3)}\left[2E_3 E_1 E_1^* + 2E_3 E_2 E_2^* + 2E_1 E_2 E_4^*\right], \tag{6.14}$$

$$P_4 = \frac{3}{2}\chi^{(3)}\left[2E_4 E_1 E_1^* + 2E_4 E_2 E_2^* + 2E_1 E_2 E_3^*\right]. \tag{6.15}$$

To solve for P_4, we must substitute the polarizations into the third order nonlinear wave equation and solve using the slowly varying envelope approximation. For A_1 we find that

$$\left(-k_1^2 + 2ik_1\frac{d}{dz'} + \frac{\varepsilon\omega^2}{c^2}\right) A_1 = \frac{-4\pi}{c^2}\frac{1}{2}\chi^{(3)}\left(|A_1|^2 + 2|A_2|^2\right) A_1, \tag{6.16}$$

where z' is along k_1. Remembering that $k_1 = \frac{\varepsilon\omega^2}{c^2}$, and that A_1 and A_2 are counter propagating, we find two similar differential equations for A_1 and A_2

$$\frac{dA_1}{dz'} = \frac{i\pi\omega}{nc}\chi^{(3)}\left(|A_1|^2 + 2|A_2|^2\right)A_1 = i\kappa_1 A_1, \quad (6.17)$$

$$\frac{dA_2}{dz'} = \frac{-i\pi\omega}{nc}\chi^{(3)}\left(|A_2|^2 + 2|A_1|^2\right)A_2 = -i\kappa_2 A_2, \quad (6.18)$$

where κ_1 and κ_2 are defined as

$$\kappa_1 = \frac{\pi\omega}{nc}\chi^{(3)}\left(|A_1|^2 + 2|A_2|^2\right), \quad (6.19)$$

$$\kappa_2 = \frac{\pi\omega}{nc}\chi^{(3)}\left(|A_2|^2 + 2|A_1|^2\right). \quad (6.20)$$

Under the plane wave assumption the phase changes as the wave propagates, but its amplitude $|A|^2$ remains constant. Thus, κ_1 and κ_2 are constants and the solutions for A_1 and A_2 are trivially complex exponentials

$$A_1(z') = A_1(0)e^{i\kappa_1 z'}, \quad (6.21)$$

$$A_2(z') = A_2(0)e^{i\kappa_2 z'}. \quad (6.22)$$

We have now determined the fields E_1 and E_2. Since we are dealing with plane waves, $|E_1|^2$ and $|E_2|^2$ are constants, so the only term that isn't constant is $E_1 E_2$. From our solution above

$$E_1 E_2 = A_1(0)A_2(0)e^{i(\kappa_1 - \kappa_2)z'}. \quad (6.23)$$

Substituting $E_1 E_2$ into the nonlinear wave equation for A_3 and A_4 yields,

$$\frac{dA_3}{dz'} = \frac{2i\pi\omega}{nc}\chi^{(3)}\left[\left(|A_1|^2 + |A_2|^2\right)A_3 + A_1 A_2 A_4^*\right] \quad (6.24)$$

$$\frac{dA_4}{dz'} = \frac{-2i\pi\omega}{nc}\chi^{(3)}\left[\left(|A_1|^2 + |A_2|^2\right)A_4 + A_1 A_2 A_3^*\right]. \quad (6.25)$$

Now if we assume $|A_1| = |A_2|$, easily attainable using one laser and a beam splitter, we find that $\kappa_1 = \kappa_2$ from Equation 6.21 and Equation 6.22, and that $A_1 A_2$ is constant. This implies automatic phase matching as $\kappa_1 - \kappa_2 \to 0$. In this case equation 6.24 and equation 6.25 become

$$\frac{dA_3}{dz} = i\kappa_3 A_3 + i\kappa A_4^*, \quad (6.26)$$

$$\frac{dA_4}{dz} = -i\kappa_3 A_3 - i\kappa A_3^*. \quad (6.27)$$

Solving equation 6.26 and equation 6.27

$$A_3'^*(L) = \frac{A_3'^*(0)}{\cos(|\kappa|L)}, \qquad (6.28)$$

$$A_4'^*(0) = \frac{i\kappa}{|\kappa|} \tan(|\kappa|L) A_3'^*(0). \qquad (6.29)$$

where A_4' is an amplified version of A_3 and L is the length the beam has traveled into the material.

Chapter 7

Appendix - Homework Solutions

List of Tables

5.1 $\chi^{(3)}$ and the response times of several mechanisms. 138

List of Figures

1.1 Kerr observed the change in transmittance through a sample between crossed polarizers due to an applied voltage. Inset at the bottom left shows the orientations of the polarizers and the applied electric field due to the static voltage. 2
1.2 (left)A system is excited by a photon if its energy matches the difference in energies between two states. (right) Two-photon absorption results when two photons, each of energy $(E_2 - E_1)/2$, are sequentially absorbed. 3
1.3 The experiment used by Franken and coworkers to demonstrate second harmonic generation. The inset shows an artistic rendition of the photograph that recorded the two beams. 4
1.4 In the Optical Kerr Effect, a strong pump laser is polarized 45° to the weak probe beam, causing a rotation of its polarization. . 5
1.5 Polarizer set perpendicular to the original wave propagation direction (the first defines the polarization of the incident wave.) . 6
1.6 Material polarized by an external electric field. 11
1.7 Plot of the electric field across the interface of a polarized material. 13
1.8 Light inducing a polarization inside a material. The light source is a monochromatic plane wave. 14
1.9 A small volume element of material in an electric field 14

2.1 Charge On A Spring 20
2.2 Harmonic Oscillator Free Body Diagram 20
2.3 N Number of Springs 21
2.4 Susceptibility Divergence 22
2.5 Harmonic Oscillator Potential 22
2.6 Real and Imaginary Susceptibility 27
2.7 Plot of the real and imaginary parts of $\chi^{(1)}$. ω_0 is the resonant frequency of the system. 30

2.8 Map of resonances of $\chi^{(2)}$ 31

2.9 Illustration of the macroscopic model that defines refractive index. Longer (red) lines represent phase fronts of the external electric field in and outside the material and the shorter (dark blue) lines represent the electric field generated by the induced dipoles fields (arrows) inside the sample. The superposition of the induced electric dipole fields and plane waves inside the sample leads to a plane wave with a decreased phase velocity, which defines the concept of the refractive index. 32

2.10 Optical Kerr Effect 43

2.11 The value of the contour integral is composed of contributions from the three major pieces of the path: (1) The semicircle of increasing radius for which $\omega_I > 0$. (2) The real axis, which avoids the singular point $\omega = \omega'$ by detouring around on a semicircular arc. (3) The pole $\omega = \omega'$ is integrated using Cauchy's integral formula. 45

3.1 Slowly varying envelope approximation: the variation of $A(z)$ with respect to z is much smaller than the wavelength of the field's oscillation. 55

3.2 Sum frequency generation: the nonlinear interaction of two incident waves of frequencies ω_1 and ω_2 with a sample of with thickness ł results in the generation of a wave of frequency $\omega_3 = \omega_1 + \omega_2$. 56

3.3 The intensity of generated light as a function of sample length, ł, is proportional to $A_1 A_2 \chi^{(2)}$ and a small-amplitude oscillating function in the non-depletion regime (solid red curve) and in the phase-matching condition (the dashed green curve), it is proportional to ł^2. 60

3.4 Sum frequency generation in the small depletion regime is represented by the nonlinear interaction of two incident waves with frequencies ω_1 and ω_2 within the material, to generate a photon of frequency ω_3. Downstream, this generated photon and a photon of frequency ω_2 interact and create a photon of frequency ω_1. The inset shows how the energy is conserved in the process of the destruction of a photon of frequency ω_3 and creation of two photons of frequencies ω_1 and ω_2. 62

3.5 The Manley-Rowe equation expresses the fact that the absolute change in the number of photons at every frequency is the same in each interaction. For example, the destruction of one photon at ω_1 and one photon at ω_2 yields one photon created at frequency ω_3. 65

3.6 Comparison of the intensities of the three beams of frequency ω_1, ω_2, and ω_3. in the sum frequency generation process. I_1 and I_2 have their minima when I_3 peaks. 66

3.7 The idler, ω_1, and the pump, ω_2, undergo sum frequency generation in a nonlinear crystal and create an output with frequency $\omega_3 = \omega_1 + \omega_2$. 67

3.8 The intensities at frequencies ω_1 and ω_3 plotted as a function of propagation distance. 69

3.9 The pump beam at frequency ω_1 and the amplified beam at frequency ω_2 undergo difference frequency generation in a nonlinear crystal. This creates the difference beam at frequency $\omega_3 = \omega_1 - \omega_2$. 70

3.10 The intensities at frequencies ω_2 and ω_3 as a function of depth into the material. 73

3.11 In second harmonic generation, two incident wave of frequency ω interact with a nonlinear medium, generating an output wave of frequency 2ω. 73

3.12 This process is difference frequency generation as we have seen in the previous section. The fundamental wave of frequency ω and the generated second harmonic wave of frequency 2ω interacts with the material and propagates through the material resulting in an output wave of frequency ω. 74

3.13 Intensities oscillate while two waves of of frequencies ω and 2ω propagate in the medium under phase-matching condition. This behavior is due to the initial non-zero incident intensities of both waves. 75

3.14 The energy is converted to the second harmonic wave from the fundamental wave. Notice that initially the slope of the fundamental wave is zero but that of the second harmonic wave begins with a non-zero value. 76

3.15 Phase-mismatching reduces the efficiency of generating second harmonic wave significantly. 76

3.16 The undepleted limit solution of coupled equations. 77

3.17 Rotating a quartz sample changes the effective thickness of the sample. Consequently, the intensity of the second harmonic generation varies with respect to the orientation of the quartz sample. 78
3.18 The intensity of second harmonic wave depends on the refractive angle ϕ, of the fundamental wave. Here we neglect the ϕ dependence of the transmittance. 78
3.19 The intensity of second harmonic generation varies with respect to the orientation of a quartz sample. 79
3.20 Phase-matching can not be found in the normal dispersion region. 80
3.21 A specific phase-matched pair of wavelengths may be found in the anomalous region, but not arbitrary wavelengths satisfy the phase-matching condition. 80
3.22 Uniaxial birefringence is characterized by an ellipsoid showing the ordinary refractive index n_o and the extraordinary refractive index n_e. 81
3.23 A plane wave travels along $\hat{k}$ direction. n_o is the component of the refractive index along the perpendicular polarization $E_\perp$, and n_e is the component along the parallel polarization $E_\parallel$. . . . 81
3.24 The dispersion of the refractive index under different orientations of a positive uniaxial crystal. Second harmonic generation can be achieved by varying the orientation of the crystal. 83
3.25 The dispersion of the refractive index under different orientations of a negative uniaxial crystal. 84

4.1 Space is equally divided into unit cells of volume L^3 86
4.2 Spontaneous emission of a photon due to de-excitation 92
4.3 Stimulated emission from deexcited system in the presence of identical photons 94
4.4 Particle trajectories of collision. (left): Inelastic collision; (right): Elastic collision. 103
4.5 Space-time diagram for above. 103
4.6 The Feynman diagram that represents the interaction between two electrons. 104
4.7 Molecular absorption of light 104
4.8 The two Feynman diagrams for the linear susceptibility 105
4.9 Example for vertex rule when photon is absorbed 106
4.10 Example for vertex rule when photon is emitted 107
4.11 Example for propagator rule to second-order in the field. 108
4.12 Feynman diagrams for sum frequency generation 109

4.13 Sum frequency generation case a 110
4.14 Outgoing photon in the propagator 110
4.15 A plot of the imaginary part of α as a function of energy is used to demonstrate the decay width. 111
4.16 a) The absorption spectrum of three different molecules with different velocities (red), and the Boltzman distribution of peak positions (dashed curve). b) Shift in the spectrum of molecules that are interacting with each other. c) A schematic picture of the superposition of molecules in an ensemble weighted by the Boltzman factor, leading to a inhomogeneous broadening. 113
4.17 a) Discrete model: the material consists of atoms or molecules. b) Continuous model: The material is assumed a uniform dielectric and a molecule inside the material is treated as a discrete entity inside a cavity. 127
4.18 An electric field applied to a material induces charges on its surfaces. A molecule in the material, modeled as being inside a dielectric cavity (small arrow), is affected by both the applied field and the polarization field. The polarization field induced in the cavity is shown in the inset. 127
4.19 If the cavity radius is much smaller than the plane wave's wavelength, the electric field will be spatially uniform inside the cavity, which is necessary for the Lorentz-Lorenz local field model. The arrows are the electric fields. 129

5.1 A pump beam of light influences the propagation of a probe beam. 133
5.2 Electronic mechanism of $\chi^{(3)}$. A molecule with a nuclear framework represented by a blue arrow, and electrons represented by the surrounding green dots. The electron cloud is deformed by the electric field, as shown above. Note that the degree of deformation is greatly exaggerated. 135
5.3 Reorientational mechanism of $\chi^{(3)}$. The nuclei of each molecule are shown as arrows. 136
5.4 Electrostriction mechanism of $\chi^{(3)}$. The effective width of the light beam is shown by the dotted lines. The green dots represent molecules in solution. 136
5.5 Saturated absorbtion mechanism of $\chi^{(3)}$. 137
5.6 Thermal mechanism of $\chi^{(3)}$. 137
5.7 An anisotropic molecule in an electric field with anisotropic polarizability. Associated with the semi-major and semi-minor axes are the polarizabilities, α_3 and α_1, respectively. 142

5.8 (a) The lowest orientational energy state; and, (b) the highest orientational energy state of an ellipsoidal anisotropic molecule in a uniform electric field. 143
5.9 The energy per anisotropic molecule in an electric field, $\vec{E}$. 144
5.10 $\frac{\delta n_{\text{linear}}}{\delta n_{\text{circular}}}$ as a function of the laser pulse width τ. 148
5.11 Axes Transformation 154
5.12 Transitions for One- and Two-Photon States 157
5.13 Transitions for One-Photon States in a Liquid 157
5.14 Wavefunctions in x and y 158

6.1 Phase Conjugate Mirror 162
6.2 Phase Conjugate Mirror 163
6.3 Phase Conjugate Mirror 163

Bibliography

[1] J. Kerr, "A new relation between electricity and light: dielectrified media birefringent," Phil. Mag. S. **50**, 337–348 (1875).

[2] J. Kerr, "Electro-optic observations on variaou liquids," Phil. Mag. **8**, 85–102,229–245 (1875).

[3] J. Kerr, "Electro-optic observations on various liquids," J. Phys. Theor. Appl. **8**, 414–418 (1879).

[4] T. H. Maiman, "Stimulated optical radiation in ruby," nature **187**, 493–494 (1960).

[5] M. Goeppert-Mayer, "Über Elementarakte mit zwei Quantensprüngen," Annalen der Physik **401**, 273–294 (1931).

[6] P. A. Franken, A. E. Hill, C. W. Peters, and G. Weinreich, "Generation of Optical Harmonics," Phys. Rev. Lett. **7**, 118–119 (1961).

[7] G. Mayer and F. Gires, "Action of an intense light beam on the refractive index of liquids," Comptes Rendus, Acad. Sci. Paris **258**, 2039 (1964).

[8] R. W. Boyd, *Nonlinear Optics* (Academic Press, 2009), 3rd edn.

[9] J. D. Jackson, *Classical Electrodynamics* (Wiley, New York, 1996), 3rd edn.

www.ingramcontent.com/pod-product-compliance
Lightning Source LLC
Chambersburg PA
CBHW051651170526
45167CB00001B/416